Library of
Davidson College

EXPERIMENTAL COELENTERATE BIOLOGY

University of Hawaii Press
Honolulu 1971

EXPER

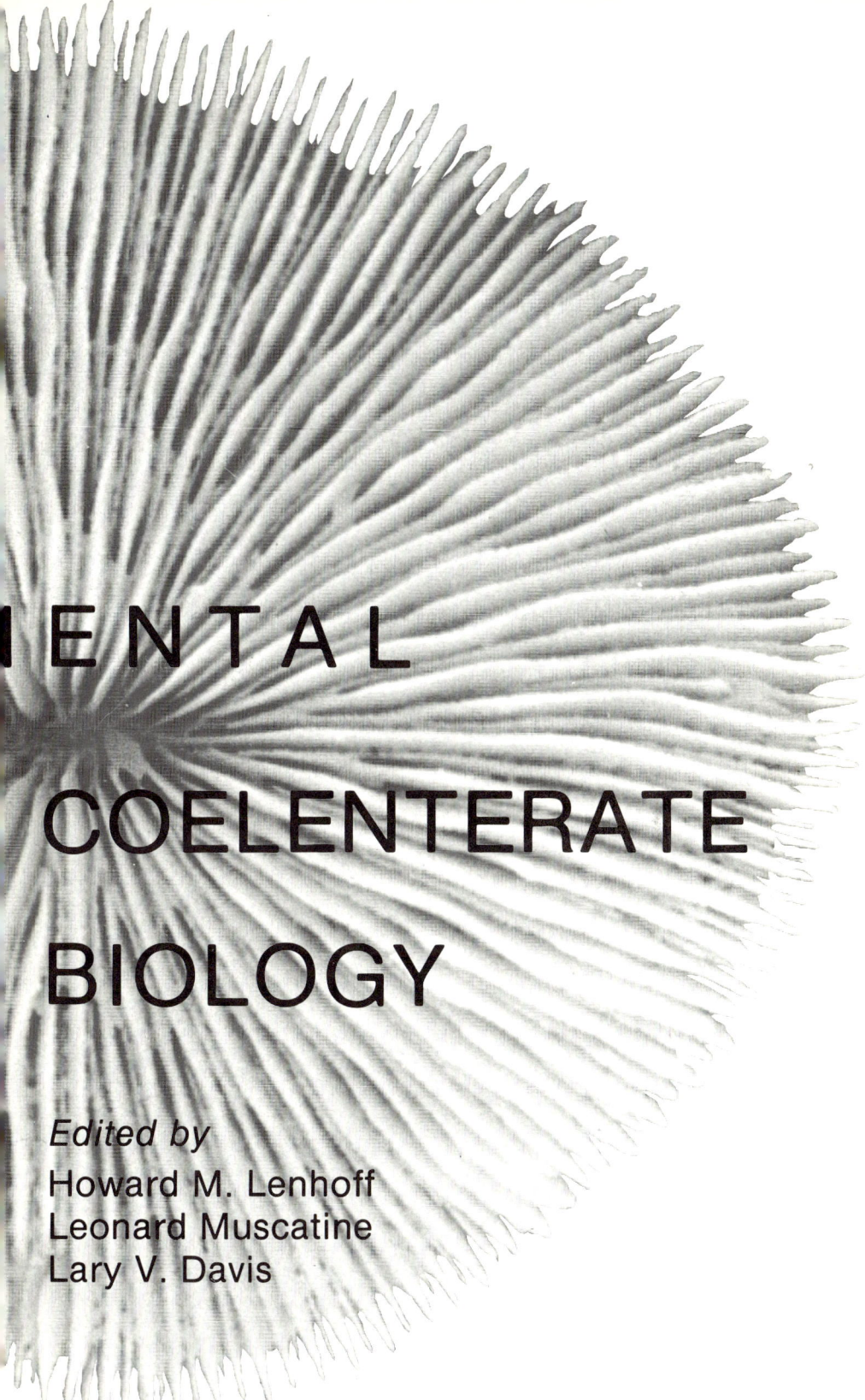

ENTAL

COELENTERATE

BIOLOGY

Edited by
Howard M. Lenhoff
Leonard Muscatine
Lary V. Davis

Library of Congress Catalog Card Number 73-127331
ISBN 0-87022-454-9
Copyright © 1971 by University of Hawaii Press
All Rights Reserved

MANUFACTURED IN THE UNITED STATES OF AMERICA

Dedication

This book is the product of a summer course in which students were trained to investigate the biology of coelenterates by using a variety of experimental procedures. The results incorporate the knowledge and experience which the instructors of the course gained from their own scientific mentors and passed on to the students. Some of these men showed us that coelenterates are not only valuable for studying a wide range of biological problems, but are interesting organisms in themselves. Those mentors who are primarily biochemists or biophysicists provided us with the insight and example of applying quantitative procedures of their particular specialty to the study of the more "classical" problems of biology.

The influence of these men in the papers presented in this volume will be apparent to those who know their work: A. A. Benson, C. Hand, N. O. Kaplan, W. F. Loomis, O. H. Lowry, and members of the Biophysics Section of the Carnegie Institution of Washington. To them this volume is gratefully dedicated.

The Editors

Contents

Foreword | PHILIP HELFRICH ix
1 Introduction | HOWARD M. LENHOFF, LEONARD
   MUSCATINE, AND LARY V. DAVIS 3

PART ONE | GROWTH AND DEVELOPMENT

2 Principles of Coelenterate Culture Methods | HOWARD
   M. LENHOFF 9
3 Growth and Development of Colonial Hydroids | LARY
   V. DAVIS 16
4 Some Common Coelenterates in Kaneohe Bay, Oahu, Hawaii |
   S. ARTHUR REED 37
5 Influence of Environmental Factors on the Growth of
   Bougainvillia sp. | JOANNE TUSOV AND LARY
   V. DAVIS 52
6 Techniques for Raising the Planula Larvae and Newly
   Settled Polyps of *Pocillopora damicornis* | S. ARTHUR
   REED 66

PART TWO | FEEDING BEHAVIOR, FOOD TRANSPORT, AND METABOLISM

7 Research on Feeding, Digestion, and Metabolism in
   Coelenterates: Some Reflections | HOWARD M. LENHOFF . 75
8 The Feeding Biology of the Gymnoblastic Hydroid
   Pennaria tiarella | ROSEVELT L. PARDY 84
9 Valine Activation of Feeding in the Sea Anemone
   Boloceroides | K. JUNE LINDSTEDT 92
10 The Chemical Control of the Feeding Behavior in Some
   Hawaiian Corals | RICHARD N. MARISCAL 100
11 Paths and Rates of Food Distribution in the Colonial
   Hydroid *Pennaria* | JOHN REES 119
12 Ingestion and Assimilation of Bacteria by Two
   Scleractinian Coral Species | L. H. DISALVO 129

13 The Formation and Assimilation of Alcohol-Soluble Proteins during Intracellular Digestion by *Hydra littoralis* and *Aiptasia* sp. | GORDON R. MURDOCK 137
14 Kinetics of Incorporation of ^{14}C-Proline into Mesogleal Protocollagen and Collagen of the Sea Anemone *Aiptasia* | JOHN M. GOSLINE 146
15 Effect of a Disulfide Reducing Agent on the Nematocyst Capsules from Some Coelenterates, with an Illustrated Key to Nematocyst Classification | RICHARD N. MARISCAL . . . 157
16 Glucose-6-Phosphate Dehydrogenase and 6-Phosphogluconate Dehydrogenase Activities in Coelenterates | DENNIS A. POWERS 169

PART THREE | ENDOSYMBIOSIS WITH ALGAE

17 Endosymbiosis of Algae and Coelenterates | LEONARD MUSCATINE 179
18 Patterns of $^{14}CO_2$ Uptake by *Chlorohydra viridissima* | ALINA M. SZMANT 192
19 Transfer of Photosynthetic Products from Symbiotic Algae to Animal Tissue in *Chlorohydra viridissima* | ERIC EISENSTADT 202
20 Uptake and Utilization of ^{14}C-Glycine by *Zoanthus* and Its Coelenteric Bacteria | AMADA REIMER 209
21 Transfer of ^{35}S-Labeled Material from Food Ingested by *Aiptasia* sp. to Its Endosymbiotic Zooxanthellae | CLAYTON B. COOK 218

PART FOUR | CALCIFICATION

22 Calcification in Corals | LEONARD MUSCATINE 227
23 Sources of Carbon in the Skeleton of the Coral *Fungia scutaria* | VICKI BUCHSBAUM PEARSE 239
24 Effects of Temperature on the Rate of ^{45}Calcium Uptake by *Pocillopora damicornis* | CONRAD CLAUSEN 246
25 Organic Matrices Associated with $CaCO_3$ Skeletons of Several Species of Hermatypic Corals | STEPHEN D. YOUNG 260

APPENDIX Two Methods for Fractionating Small Amounts of Radioactive Tissue | HOWARD M. LENHOFF AND BERTON ROFFMAN 265

Foreword

This volume is a result of research done by the editors, together with a select group of students, during the summer of 1967 at the Hawaii Institute of Marine Biology on Coconut Island, Oahu, Hawaii. In an effort to extend the influence of the University of Hawaii as an institution striving for excellence, members of the Institute conceived a unique program which made maximum use of limited laboratory facilities in the tropical environs of Hawaii, and, at the same time, made a contribution to graduate education in marine biology. A departure was made from the usual pattern of traditional courses and directed research. Informal group-research training was initiated with the guidance of outstanding scholars, each of whom had a different approach to some aspect of coelenterate biology. Through the foresight and generous support of the National Science Foundation, this program came to fruition.

The program was modest; the atmosphere was scholarly, dynamic, and creative. It was the first exposure for many of the students to a complex coral reef community. New techniques and approaches were introduced which allowed the students to trace biochemical pathways, elucidate vital processes, and explain ecological relationships in a variety of coelenterates. The instructor-student relationship was highly conducive to the production of ideas that evolved into fruitful research.

Most of the research described in this volume was accomplished during a 12-week period. Interests developed and nurtured during that summer have persisted, and we now know that this endeavor has strongly influenced and given direction to many of the participants. This volume stands as testimony that we can make significant contributions to graduate training by exposing our students to a rich tropical biota, and, at the same time, advance our own knowledge of a particular discipline.

It was my distinct privilege to have been associated with the biologists whose concerted efforts produced this book. It is my hope that their work may serve as an example of a unique and rewarding approach to graduate education.

Philip Helfrich   July 1, 1971

EXPERIMENTAL COELENTERATE BIOLOGY

HOWARD M. LENHOFF
University of California at Irvine
LEONARD MUSCATINE
University of California at Los Angeles
LARY V. DAVIS
Commission on Undergraduate Education in the Biological Sciences, Washington, D.C.

CHAPTER 1

Introduction

A pilot program to train graduate students for experimental research with marine animals was initiated at the Hawaii Institute of Marine Biology during the summer of 1967. A group of 15 graduate students and five instructors spent 3 months using some relatively simple biochemical techniques to investigate animals of a single phylum, the Coelenterata. Rather than concentrating on one aspect of coelenterate biology, the group investigated such diverse areas as collagen biosynthesis, chemoreception, symbiosis, calcification, carbohydrate metabolism, protein digestion and utilization, and ecology of reef corals.

The program proved successful. All the students made significant progress in research; their results are published in this volume. Eight of the papers have already appeared in modified form in scientific journals. Twelve of the students continued to work on the same or closely related problems for their Ph.D. theses and/or postdoctoral research. The general format of the program was followed by at least two other programs: one, on molluscs, at the Hawaii Institute of Marine Biology in 1968, and a second, again on coelenterates, at the Marine Laboratory of Hebrew University in Eilat, Israel in 1970.

We can attribute the success of this course to a happy combination of many factors. Because our experience may be valuable to others planning similar programs, we discuss each factor in some detail.

FORMAT OF THE COURSE

The primary goal was to combine field and laboratory experience so that each student would initiate a research problem that dealt experimentally with the biology of an organism. Thus, the first week was devoted mostly to field trips, to collecting and identifying specimens, and to observing the organisms both in the field and in the laboratory.

Lectures were given during the 2nd week describing previous investigations on hydras and some marine coelenterates, and the experimental techniques which have been developed were demonstrated in the student laboratories.

By the 3rd week most students had selected a specific research problem on which they were expected to make significant progress in the ensuing 2 months. After that point, the students and faculty were engrossed in their research. Every day at least 1 to 2 hours were spent in seminars in which the research in progress was examined critically by all the participants.

During the last 2 weeks the students wrote manuscripts which were later prepared for journal publication and for this volume. The course was climaxed by a 2-day colloquium, open to the public, at which each student presented the results of his research project.

USE OF A LABORATORY-GROWN ANIMAL AS A MODEL

We instructors in the course had been using the freshwater hydras at our home institutions because of the availability of techniques for raising and maintaining these animals in large numbers under controlled conditions. Hence, we had gathered a great deal of information about these simple coelenterates—information which provided a sturdy foundation for virtually all of the projects carried out by the students on the marine coelenterates. Thus, at this marine station in the center of the Pacific Ocean, considerable emphasis was placed on learning about the biology of the freshwater hydras.

Our use of hydra points out one of the major advantages in using a laboratory-grown organism as an experimental prototype for research in marine biology. Except for the fortunate few who are at institutions close to the marine environment, many marine biologists can conduct research in the field for only 2 or 3 months a year. But, by having available year-round a laboratory-grown animal that can serve as a prototype for experimentation on marine species, those 2 or 3 months can be extremely productive. With the laboratory-grown organism, problems can be defined and techniques can be developed. These can be pursued more intensively with little delay during the typical 3-month stay at a marine station. For example, information obtained from the freshwater hydra on nematocysts, chemical control of feeding, endosymbiotic interactions, and intracellular digestion proved invaluable for investigating similar structures and phenomena in marine coelenterates. The work reported here by the students affords ample proof of this.

USE OF A MULTIDISCIPLINARY APPROACH

Marine biology as studied today covers an extremely broad area. For this program it was decided to restrict the activities of the participants to a single phylum because we did not have a large enough faculty to cover the large variety of species encountered at a marine station. This "restriction" to a single phylum did not signify that a narrow approach would be taken. On the contrary, the instructors, though all experienced with the coelenterates, brought a number of different disciplines to bear upon a common set of problems. Observing this effective multidisciplinary approach in action, the student could not help but be caught up in the excitement and enthusiasm of the senior and guest faculty.

MYSTIQUE OF THE MARINE LABORATORY

If you ask any biologist what his recollections are of his first summer at a marine station, he will probably tell you that it was one of the most memorable experiences of his student days. Such was also the case with us at the Hawaii Institute of Marine Biology in the summer of 1967.

It is not difficult to understand this "mystique." Most summer students come to a marine station from land-bound institutions after years of studying organisms from textbooks and odoriferous formaldehyde jars. Suddenly these students are thrust into a situation where they can witness hundreds and hundreds of living species in their natural environment. Their book-gained knowledge becomes real. The diversity and beauty of living organisms exceed their every expectation.

On top of this is the pleasant fact that a good many of the marine stations are located in beautiful, colorful, and somewhat isolated locales. The romantic subtropical site of the Hawaii Institute of Marine Biology, with its distinctive local atmosphere, certainly contributed to the happy attitude of the students and to the eventual success of the course. Among the most memorable events were the guest lectures on the geology and history of Hawaii and the performances by Kaupena Wong of the Bishop Museum. Through Mr. Wong we were privileged to taste the rich musical and cultural heritage of the Hawaiian people.

THE "COMPLEAT" RESEARCH EXPERIENCE

A summer program at a marine station, if carefully designed, can be a memorable experience in a graduate student's career, especially if he attends such a program after successfully completing most of his course work toward the Ph.D. degree. Imagine—after 17-19 years of formal study, the student is declared ready to be initiated into the rites of independent research. He is told that his only responsibility is to produce a creditable piece of original research that will be published in a book and possibly a journal.

Our students responded. Within a few weeks they found their organisms and problems and moved from field studies to the laboratory.

After 2 months of research, they wrote the initial drafts of their first scientific papers and presented their results to a scientific audience. Field studies, laboratory research, oral reports, and publication—the complete research experience in 3 months.

The words of the day were research—and the book. Their results are in the following pages.

PART 1

GROWTH AND DEVELOPMENT

HOWARD M. LENHOFF CHAPTER 2
University of California at Irvine

Principles of Coelenterate Culture Methods

The greatest stimulus to the current revival in coelenterate research was the development by Loomis (1953, 1954) of methods by which hydra could be cultured in the laboratory under controlled conditions. Those methods made it possible for biologists in virtually every part of the world to take these hardy invertebrates from neighboring ponds and maintain them in the laboratory. Soon after the first publication of Loomis's methods, there came reports on the mass culture of hydra (Loomis and Lenhoff 1956), and on the successful laboratory culture of the brackish-water colonial gymnoblastic hydroid *Cordylophora lacustris* (Fulton 1960, 1962), the marine colonial hydroid *Podocoryne carnea* (Braverman 1962*a, b*), the scyphozoan *Aurelia aurita* (Spangenberg 1965), and others (see Lenhoff 1968).

GENERAL CONSIDERATIONS

These advances in coelenterate husbandry have had two broad consequences. First, much control over the physiology of the animal is now possible. For example, by merely controlling such variables as feeding schedule, composition of ionic environment, and pH, it is possible to obtain large numbers of animals that are in the same developmental stage and that respond nearly in synchrony to the hydra feeding activator glutathione. Sexual differentiation can be controlled in a predictable fashion in *Hydra littoralis* by partial pressure of carbon dioxide (Loomis 1957) or in *Chlorohydra viridissima* by the degree of feeding (Rutherford, Hessinger, and Lenhoff 1965). Calcium ions can control nematocyst discharge (Lenhoff and Bovaird 1959), and cyanide can inhibit the feeding response but not the discharge of nematocysts (Brown, Reasor, and Lenhoff, unpublished). A calculated treatment with cesium and sodium ions can produce hypersensitive animals (Lenhoff 1966). Manipulation of the concentrations of environmental sodium ions can affect cnidocyte migration, and can even produce developmental monsters (Lenhoff and Bovaird 1960).

The specific ionic factors affecting the growth and maintenance of hydras have been particularly well studied. Both calcium ions (Loomis 1954) and sodium ions (Lenhoff and Bovaird 1960; Muscatine and Lenhoff 1965) are indispensable to *Hydra littoralis* and *Chlorohydra viridissima*. In the absence of Ca^{++} the animals disintegrate in several hours (Lenhoff 1968). *Hydra littoralis* may survive in a sodium-free solution, but such deprivation gives rise to gross abnormalities and to cessation of growth. Specimens of *Chlorohydra viridissima* disintegrate within several days if Na^+ is omitted from the environment. The symbiotic strain has a greater tolerance to Na^+ deprivation than the aposymbiotic strain (Lenhoff and Bovaird 1960). Muscatine and Lenhoff (1965) demonstrated that Mg^{++} and K^+, though not absolutely necessary, enhance the growth rate of *C. viridissima*. Potassium ions, but not magnesium ions, have a similar effect upon *Hydra littoralis* (Lenhoff 1966). In both species K^+ also causes a two- to three-fold increase in tentacle length (Lenhoff 1966). The indispensability of environmental K^+, in addition to Ca^{++} and Na^+ and the growth-enhancing effects of Mg^{++}, has been demonstrated for the brackish-water hydroid *Cordylophora lacustris* (Fulton 1960, 1962). Environmental Ca^{++} is also required for nematocyst discharge and for activation of the feeding response in these three species (Lenhoff and Bovaird 1959; Fulton 1963). No specific anion requirements have been found for hydras. *Hydra littoralis* grows equally well with either nitrate or chloride ions (Lenhoff, unpublished observations). In contrast, Fulton (1962) found that, though *Cordylophora lacustris* can survive and feed in the absence of chloride ions, it develops no new hydranths.

The second broad consequence of advances in coelenterate husbandry is perhaps the most significant. Contradictory as it may seem, the investigator has a much greater chance to uncover the natural history of an animal when he grows that animal in a laboratory culture than he would have by studying the animal only in its natural habitat. A laboratory culture enables him to observe the animal for longer periods, under more comfortable conditions, and under a greater variety of environmental and physiological circumstances. He can simulate many natural conditions and, more important, he can alter environmental conditions at will.

It is by such surveillance of our cultures that we have gleaned knowledge of (*a*) hydras' growth requirements (Loomis 1954) and metabolic rate under different environmental conditions (Lenhoff and Loomis 1957); (*b*) the chemistry and physiology of hydras' defense mechanisms (Lenhoff, Kline, and Hurley 1957; Blanquet and Lenhoff 1966); (*c*) the chemical control of hydras' feeding behavior (Loomis 1955; Lenhoff 1961*a*), and other behavioral modifications (Blanquet and Lenhoff 1968); (*d*) the biochemistry of intracellular digestion (Lenhoff 1961*b*); (*e*) mechanisms of alga-hydra symbiosis (Muscatine and Lenhoff

1963); (*f*) such developmental features as cnidocyte migration (Lenhoff 1959; Lenhoff and Bovaird 1961), and the changes which occur during budding (Li and Lenhoff 1961); and (*g*) genetic developmental abnormalities (Lenhoff 1965*a*).

The truism that many of our best experiments have come from some accidental observation is especially applicable in the case of laboratory-grown organisms. Because opportunities for continued observation and experimentation on an intact organism are greatly increased, a wide variety of unusual situations are chanced upon. Accordingly, the findings listed in the preceding paragraph, which may appear to depict an "orderly" study of various phases of the natural history of hydras, did not take place as a preconceived series of investigations. Quite the contrary. One study led to another through circuitous routes. For example, investigations on chemical control of feeding (Lenhoff 1961*a*) led to discovery of the developmental mutants (Lenhoff 1965*a*). Investigations of these mutants led to discovery of control of gonadogenesis in the green hydra *Chlorohydra viridissima* (Rutherford, Hessinger, and Lenhoff 1965). Studies of gonadogenesis led directly to discovery of tyrosine control of neck formation (Blanquet and Lenhoff 1968).

The advantages of such a laboratory study of the natural history of an animal may not be realized if the investigator raises the organism solely to solve a particular problem. The investigator must, instead, be tuned to the breadth of biology. As his work dictates, he should be prepared to venture from one discipline of biology (such as development, behavior, or biochemistry) into another, without being held back by fear of making a few mistakes. By such an organismic approach he will gain more than a knowledge of the particular problem under immediate study; the information he acquires regarding almost any aspect of an organism eventually will be useful in understanding other aspects (including those connected with his original problem) and, of course, in understanding the whole organism. But perhaps the greatest value—and thrill—of such a broad organismic study is that previously unsuspected phenomena may be revealed.

WHY COELENTERATES?

Among the invertebrates, the laboratory-rearing of coelenterates has been particularly successful. To what can we attribute this remarkable success in coelenterate husbandry when so little has been attained with aquatic invertebrates of other phyla (except Protozoa and some helminth parasites)? I think the following four factors are of importance: (*a*) a suitable culture solution has been devised; (*b*) a supply of live food of high and stable nutritional value is available on demand; (*c*) the animals attach to a solid surface; (*d*) the animals multiply by asexual reproduction.

Culture Solution

Because the culture solution is the immediate environment from which the animal obtains its food and into which it excretes its wastes, it is essential that this solution be changed frequently. To do this, one must be in a position to prepare an ample supply of culture solution with minimum effort.

It is especially important to control the composition of the culture solution when working with animals, such as hydras, which have a significant portion of their cells exposed directly to the environment. Hydras, because they have a certain combination of features, are unique among members of the animal kingdom: they are diploblastic, possess essentially no internal extracellular fluids (other than the contents of the gastrovascular cavity—a solution of varying composition), and live in freshwater. Most other freshwater metazoans, with the possible exception of *Craspedacusta* and the brackish-water *Cordylophora*, contain considerably more internal extracellular fluids. Thus, hydras are particularly sensitive to changes in the composition of their culture solution. I have reviewed elsewhere (Lenhoff 1965*b*, 1968) some of the effects of slight changes in the composition of the culture solution on the behavior and development of these animals.

Food

Ideally, invertebrates grown in the laboratory should be supplied with a defined medium of nutrients. Unfortunately, such a medium has not yet been found for hydras. These animals are relatively inefficient in taking up water-soluble materials, and they are contaminated with microorganisms that foul the nutrient medium. (We have recently been able to obtain some germ-free *Hydra littoralis*, but have not yet been able to culture them successfully on a mass scale.) Nauplii hatched from cysts of the brine shrimp, *Artemia salina*, have proved to be a reliable and suitable food source. They are inexpensive, readily available in large numbers, can be stored in the dry form until needed, and are a nutritionally stable and complete diet for many coelenterates. Hydras, which are carnivorous, have mechanisms for capturing and ingesting the nauplii. In the laboratory we successfully fed *Artemia* nauplii to hydras, sea anemones, hydroids, corals, medusae, and marine and freshwater flatworms. *Artemia* offers another advantage as a food source: because the nauplii can be readily obtained free of microorganisms, the amount of bacterial and fungal growth in the culture solution is significantly lowered.

Attachment to a Surface

The twice-daily regimen of changing the culture solution bathing the animals is greatly simplified if they naturally attach to a solid substratum, because (as in the case with hydras and sea anemones) the used culture solution, with its waste material, dead nauplii, etc., can be removed by merely inverting the culture tray over a sink. It is also possible to grow some coelenterates on plexiglass plates suspended vertically from racks into containers of culture solution. In this case cleaning the solution of uningested nauplii consists of simply transferring the rack with attached plates to a container of clean culture solution. Colonial hydroids can be made to attach to a surface by tying a piece of stolon with a hydranth to a glass microscope slide (see Fulton 1960). Although invertebrates that do not adhere firmly to the container can be raised in mass culture, it is relatively tedious to do so.

Asexual Reproduction

Asexual reproduction, common among coelenterates, affords a rapid means by which an organism can fill a new ecological niche (Mayr 1963), whether in nature or in the laboratory. There is a special advantage to using experimental animals which can be made to reproduce solely asexually—all of the progeny are genetically identical, except for the possible accumulation of somatic mutations.

PROPOSED PROTOCOL FOR CULTURING COELENTERATES

When attempting to culture under defined conditions a coelenterate never before raised in the laboratory, the researcher is more likely to attain success using animals which adhere to a solid surface, reproduce asexually, and feed on small crustaceans like *Artemia* nauplii.

Once a suitable culture solution has been developed, it is advantageous to devise a simple measure—wet weight, number of hydranths, number of individuals, or stolon length—by which growth and reproduction of the organisms may be quantitatively measured. The effects of variations in such environmental factors as pH, oxygen, salinity, and temperature on the animals may then be accurately determined. How does the frequency and amount of feeding affect growth? What about light-sensitivity? Does the animal undergo any biological rhythms? After those factors have been analyzed, a "plateau" set of conditions can be selected where growth will not be affected greatly by slight variations in such factors as pH, oxygen, or osmotic pressure. Under these plateau conditions, and at a temperature predetermined to be safe, the principal variable that will affect growth will be the amount of food ingested by the animal. Only after such parameters are examined and defined will there be a solid basis for further experimentation on that organism.

LITERATURE CITED

Blanquet, R. S., and H. M. Lenhoff. 1966. A disulfide-linked collagenous protein of nematocyst capsules. *Science* **154**: 152–153.

———. 1968. Tyrosine enteroreceptor of hydra: Its function in eliciting a behavior modification. *Science* **159**: 633–634.

Braverman, M. H. 1962a. Studies in hydroid differentiation. I. *Podocoryne carnea* culture methods and carbon dioxide induced sexuality. *Experimental Cell Research* **27**: 301–306.

———. 1962b. *Podocoryne carnea*, a reliable differentiating system. *Science* **135**: 310–311.

Fulton, C. 1960. Culture of a colonial hydroid under controlled conditions. *Science* **132**: 473–474.

———. 1962. Environmental factors affecting growth of *Cordylophora*. *J. Experimental Zoology* **151**: 61–78.

———. 1963. Proline control of the feeding reaction of *Cordylophora*. *J. General Physiology* **46**: 823–837.

Lenhoff, H. M. 1959. Migration of ^{14}C-labeled cnidoblasts. *Experimental Cell Research* **17**: 570–573.

———. 1961a. Activation of the feeding reflex in *Hydra littoralis*. I. Role played by reduced glutathione, and quantitative assay of the feeding reflex. *J. General Physiology* **45**: 331–334.

———. 1961b. Digestion of protein in *Hydra* as studied using radioautography and fractionation by differential solubilities. *Experimental Cell Research* **23**: 335–353.

———. 1965a. Cellular segregation and heterocytic dominance in hydra. *Science* **148**: 1105–1107.

———. 1965b. Some physicochemical aspects of the macro- and micro-environments surrounding hydra during activation of their feeding behavior. *American Zoologist* **5**: 515–524.

———. 1966. Influence of monovalent cations on the growth of *Hydra littoralis*. *J. Experimental Zoology* **163**: 151–156.

———. 1968. Chemical perspectives on the feeding response, digestion, and nutrition of selected coelenterates. In *Chemical zoology,* vol. 2, M. Florkin and B. Scheer, eds., pp. 157–221. New York: Academic Press.

Lenhoff, H. M., and J. H. Bovaird. 1959. Requirement of bound calcium for the action of surface chemoreceptors. *Science* **130**: 1474–1476.

———. 1960. The requirement of trace amounts of environmental sodium for the growth and development of *Hydra*. *Experimental Cell Research* **20**: 384–394.

———. 1961. A quantitative chemical approach to problems of nematocyst distribution and replacement in *Hydra*. *Developmental Biology* **3**: 227–240.

Lenhoff, H. M., E. S. Kline, and R. E. Hurley. 1957. An hydroxyproline-rich, intracellular, collagen-like protein of *Hydra* nematocysts. *Biochimica et Biophysica Acta* **26**: 204.

Lenhoff, H. M., and W. F. Loomis. 1957. Environmental factors controlling respiration in hydra. *J. Experimental Zoology* **134**: 171–182.

Li, Y.-Y., and H. M. Lenhoff. 1961. Nucleic acid and protein changes in budding *Hydra littoralis*. In *The biology of hydra and of some other coelenterates: 1961*, H. M. Lenhoff and W. F. Loomis, eds., pp. 441–448. Coral Gables: University of Miami Press.

Loomis, W. F. 1953. The cultivation of *Hydra* under controlled conditions. *Science* **117**: 565–566.

———. 1954. Environmental factors controlling growth in hydra. *J. Experimental Zoology* **126**: 223–234.

———. 1955. Glutathione control of the specific feeding reactions of hydra. *Annals, New York Academy of Sciences* **62**: 209–228.

———. 1957. Sexual differentiation in hydra: Control by carbon dioxide tension. *Science* **126**: 735–739.

Loomis, W. F., and H. M. Lenhoff. 1956. Growth and sexual differentiation of hydra in mass culture. *J. Experimental Zoology* **132**: 555–574.

Mayr, E. 1963. *Animal species and evolution*. Cambridge: Harvard University Press.

Muscatine, L., and H. M. Lenhoff. 1963. Symbiosis: On the role of algae symbiotic with hydra. *Science* **142**: 956–958.

———. 1965. Symbiosis of hydra and algae. I. Effects of some environmental cations on growth of symbiotic and aposymbiotic hydra. *Biological Bulletin, Woods Hole* **128**: 415–424.

Rutherford, C. L., D. Hessinger, and H. M. Lenhoff. 1965. Induction of rhythmic differentiation of ovaries in *Chlorohydra viridissima*. *American Zoologist* **5**: 722.

Spangenberg, D. B. 1965. Cultivation of the life stages of *Aurelia aurita*. *Trans., American Microscopical Society* **83**: 448–455.

LARY V. DAVIS
CHAPTER 3
Commission on Undergraduate Education in the Biological Sciences, Washington, D.C.

Growth and Development of Colonial Hydroids

In this chapter, attention will be focused on certain aspects of hydroid growth. These are (*a*) a survey of the methods which have been developed for maintaining colonial hydroids in laboratory cultures; (*b*) various quantitative expressions of growth, and the manner in which these may be influenced by either intrinsic (genetic) or extrinsic (environmental) factors; and (*c*) morphogenetic activities during stolonal elongation. For information on other aspects of the growth and development of colonial hydroids, see any of the several review articles which have appeared in the past decade (e.g., Berrill 1961; Crowell 1961; Strehler 1961; Tardent 1963, 1965).

CULTURE METHODS

General Considerations

There are many advantages in working with colonial hydroids grown in the laboratory. For example, laboratory cultures enable the investigator to control properly various environmental factors in the development of the hydroid. Also, in hydroid systematics, the correct association between the polyp and medusa stages of the same species frequently becomes possible only after both are reared in the laboratory (Brinckmann 1962; Nagao 1964; Rees 1938, 1939; Russell and Rees 1936).

Of considerably greater significance to the investigator, however, is the fact that the maintenance of colonial hydroids in laboratory cultures provides an abundant, year-round supply for developmental and nutritional experiments. It is not unreasonable to suggest that, with the exception of certain ecological relationships, virtually every aspect of the biology of colonial hydroids can be studied to considerable advantage on animals reared in the laboratory.

Browne's Work with Colonial Hydroids

Beginning with the studies of E. T. Browne (1897, 1907), several methods have been introduced for growing colonial hydroids in the laboratory. The relative ease with which hydroids can be cultured is reflected in the number of species which may be maintained in the laboratory by one or more of these methods (Table 1).

Browne's early attempts to grow colonial hydroids, however, were only partially successful. Although he was able to maintain *Syncoryne exima* and *Bougainvillia* sp. in the laboratory, he could do so only for relatively short periods of time; the colonies usually regressed and eventually disintegrated. He ascribed their death to one of two factors: (*a*) fouling of the medium from occasional sudden increases in the numbers of diatoms or other species of phytoplankton, or (*b*) lack of food caused by a decrease in zooplankton in waters near the laboratory. Although the first of these limiting factors could be controlled by frequently changing the filtered seawater in which the colonies were grown, it was difficult to provide an adequate amount of food throughout the year.

TABLE 1   Colonial hydroids maintained in laboratory cultures

Species[a]	References[b]
Anthomedusae (=Gymnoblastea)	
Acaulis ilonae	Brinckmann-Voss 1966
Amphinema dinema	Rees and Russell 1937
Amphinema rugosum	Rees and Russell 1937
Bougainvillia sp.	Tusov and Davis 1969
Bougainvillia carolinensis	Brock and Strehler 1963
Bougainvillia muscus	Browne 1907
Cladonema radiatum	Weiler-Stolt 1960
Clava multicornis	Kinne and Paffenhöfer 1965
Cordylophora lacustris	Roch 1924; Kinne 1956; Crowell 1957; Fulton 1960
Coryne tubulosa	Rees 1941
Dipurena halterata	Rees 1939
Eleutheria dichotoma	Weiler-Stolt 1960
Hydractinia echinata	Hauenschild and Kanellis 1953; Müller 1961, 1964; Toth 1965
Margelopsis haeckeli	Werner 1955
Merga galleri	Brinckmann 1962
Nemopsis dofleini	Nagao 1964
Podocoryne carnea	Braverman 1962
Podocoryne hartlaubi	Yamada 1967
Protohydra leuckarti	Muus 1966
Rathkea octopunctata	Rees and Russell 1937
Rhizorhagium album	Rees 1938
Sarsia tubulosa	Kukinuma 1966a
Stauridiosarsia japonica	Nagao 1962
Staurocladia portmanni	Brinckmann 1964
Staurocoryne filiformis	Rees 1936
Syncoryne exima	Browne 1907

TABLE 1—continued

Species[a]	References[b]
Tubularia crocea	Mackie 1966
Zanclea implexa	Russell and Rees 1936
Leptomedusae (=Calyptoblastea)	
Aequorea coerulescens, or	
"Campanulina type"	Kukinuma 1966b
Campanularia calceolifera	Miller 1966
Campanularia flexuosa	Crowell 1953
Campanularia johnstoni	Weiler-Stolt 1960
Clytia johnstoni	Hale 1960; Brock and Strehler 1963
Lovenella (=Eucheilota) clausa	Russell 1936
Mitrocomella (=Cuspidella) brownei	Rees and Russell 1937
Obelia sp.	Palincsar and Palincsar 1960
Trachylina	
Craspedacusta sowerbii	Reisinger 1957; McClary 1959; Lytle 1961; Matthews 1966

[a]Limited to published accounts of species which were maintained in laboratory cultures for a period of time during which they were fed and cared for.
[b]The references listed are those which describe the culture methods used for each species. Accounts in which established methods were employed are not included unless they were used to culture a different species. When more than one reference is listed for a species, each reference describes a different method which was employed successfully for that species.

Despite the difficulties he encountered, Browne was able to obtain excellent growth in his hydroid cultures. For example, the rate of growth that he obtained for *Syncoryne exima* is one of the highest recorded for any colonial hydroid to date (see Table 2).

Browne designed two devices in which he grew colonies of hydroids. The first of these (Figure 1), referred to as a "bell jar with plunger," was designed primarily for keeping medusae and other small floating forms alive, and the second, his "continuous current tube" (Figure 2), was built specifically for growing hydroid colonies. The major drawback to both of these devices is that they require relatively large amounts of laboratory space while providing adequate conditions for only a limited number of hydroid colonies. The significance of Browne's studies is not related so much to the particular types of apparatus he devised for culturing hydroids as it is to his delineation of the basic requirements which must be met in order to culture these animals. These requirements are (*a*) a nutritionally adequate food supply, available at all times; (*b*) a suitable medium; and (*c*) some means of preventing the depletion of required substances or the accumulation of deleterious substances, either in localized areas of the culture or throughout the medium.

TABLE 2 Growth rates of colonial hydroids

Species	Rate of Growth of Horizontal Stolons	Vertical Stems	Lateral Branches	Rate of Increase in Number of Hydranths (k)	Reference
Bougainvillia sp.	4.6 mm/day	—	—	0.32	Tusov and Davis (this volume)
Campanularia flexuosa	4.5–5.0 mm/day	—	—	0.55[a]	Crowell 1957; Wyttenbach 1968
Clava multicornis	Exponential over 40-day period $k = 0.045$ at 17° C	—	—	—	Kinne and Paffenhöfer 1966
Clytia johnstoni	3.1 mm/day	—	—	—	Hale 1964
Cordylophora lacustris	3–4 mm/day	1.5–2.0 mm/day	0.5 mm/day	0.21	Fulton 1962, 1963a
Obelia geniculata	1.0 mm/day at 10°–12° C[b] 10 mm/day at 16°–17° C 15–20 mm/day at 20° C	—	—	—	Berrill 1949a
Podocoryne carnea	Exponential over 20-day period $k = 0.23$ (approx.) (total stolon growth)	—	—	0.23 (approx.)	Braverman 1963
Protohydra leucarti	—	—	—	0.173[a]	Muus 1966
Sertularia argenta	—	1.7 mm/day	—	—	Hancock, Drinnan, and Harris 1956
Syncoryne exima	7.0 mm/day[a]	—	—	0.346[a]	Browne 1907
Tubularia crocea	1.0 mm/day	1.0 mm/day	—	—	Mackie 1966

[a] Calculated from the author's data.
[b] Free stolons.

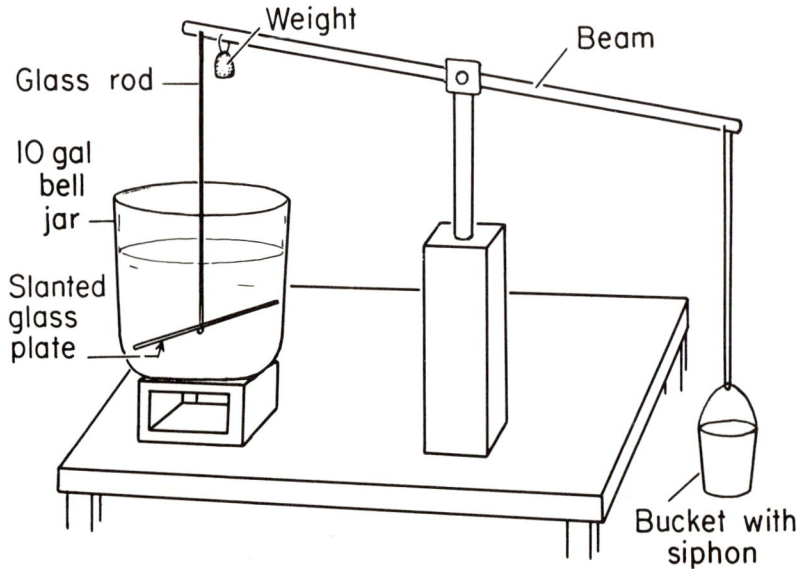

FIG. 1. "Bell jar with plunger" device for maintaining small marine organisms in the laboratory.

Nutrition

Crowell (1953) using *Campanularia flexuosa*, Hauenschild and Kanellis (1953) using *Hydractinia echinata*, and Loomis (1953) using *Hydra littoralis* were the first to demonstrate that hydroid polyps remain in good physiological condition and reproduce normally when fed on the nauplii of the brine shrimp, *Artemia salina*. These nauplii can be hatched in the laboratory from the dry, encysted, dormant gastrulae, which are always commercially available.

In most cases, only freshly hatched nauplii are fed to the polyps. In the case of *Hydractinia echinata*, however, Hauenschild (1954) found that polyps fed on newly hatched nauplii exhibited several abnormalities in growth. These abnormalities did not develop when the polyps were fed on two- or three-day-old nauplii.

Rather than relying solely on *Artemia*, several investigators have supplemented the diet with other materials. For example, Kinne (1958) reared *Cordylophora lacustris* on *Artemia* plus various annelids (*Pachydrilus lineatus, Enchytraeus albidus, Lumbricus* sp.) and Müller (1964) raised *Hydractinia echinata* on three-day-old *Artemia* nauplii, bits of fish flesh, harpacticoid copepods, and any other available zooplankton. These authors did not indicate to what extent the growth and well-being of these hydroids is dependent upon the use of such dietary supplements. For

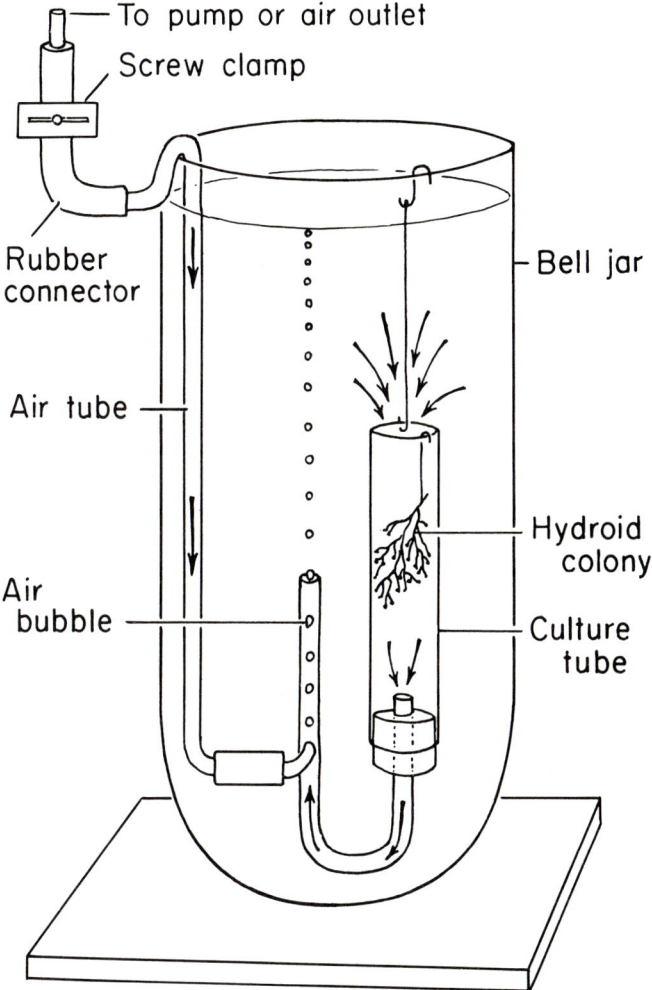

FIG. 2. "Continuous current tube" device for maintaining small sessile marine organisms.

Cordylophora, at least, the requirement for the supplement cannot be absolute, because this species grows very well when fed on *Artemia* nauplii only (Fulton 1962), as do many other species for which such supplements have never been used.

In contrast to the attached polypoid stages, hydromedusae do not grow well on a diet of *Artemia* nauplii alone. In many instances, as with *Podocoryne hartlaubi* (Yamada 1961), *Nemopsis dofleini* (Nagao 1964), and *Staurocladia portmanni* (Brinckmann 1964), the medusae can be grown only when their diet of *Artemia* is supplemented with pieces of the hepatopancreas (or other viscera) from *Mytilus* or other molluscs. In another study, Brinckmann-Voss (1966) found that, although the young medusae of *Acaulis ilonae* could be grown on *Artemia*, mature medusae grew better and produced larger gonads when *Sagitta* (a chaetognath) and various copepods were added to the diet.

Not unexpectedly, the growth of colonial hydroids is influenced by the quantity as well as the quality of the food they receive. In the gymnoblastic species *Cordylophora lacustris* (Fulton 1962; Brock and Strehler 1963), *Podocoryne carnea* (Braverman 1962), *Bougainvillia carolinensis* (Brock and Strehler 1963), and *Bougainvillia* sp. (Tusov and Davis, this volume), maximum rates of asexual budding were obtained when cultures were fed to repletion once each day. Similar results have been obtained with *Hydra littoralis* (Loomis 1954) and *Chlorohydra viridissima* (Muscatine and Lenhoff 1965). In all cases, as the interval between feedings increased, the growth rates decreased.

Contrary to these results, however, Brinckmann (1964) found that when fed to repletion daily, the polyps of *Staurocladia portmanni* disintegrated within a few days. Best results were obtained when each polyp was fed a single *Artemia* nauplius every 2nd or 3rd day. Preliminary observations on *Corymorpha palma* suggest that this species may be affected similarly by daily feedings to repletion on *Artemia* nauplii (Davis, unpublished observations).

Two species of calyptoblastic hydroids, *Campanularia flexuosa* (Crowell 1957; Brock and Strehler 1963) and *Clytia johnstoni* (Brock and Strehler 1963; Hale 1964), grew at maximum rates when fed to repletion twice each day. The hydranths of calyptoblastic hydroids are, in general, somewhat smaller than those of most gymnoblastic species, and, therefore, cannot ingest as many *Artemia* at one time as can the latter. This may account for the more frequent feedings required in calyptoblastic species.

Media

Most investigators who have attempted to rear colonial marine hydroids in the laboratory have begun with the premise that the artifical conditions should simulate, within limits, the natural environment from which the organism was obtained. Consequently, most species have been grown in (filtered) natural seawater. Artificial seawater solutions have also been used occasionally, either because natural seawater was not available or because a medium of more constant composition than seawater was desired. In recent years, "Instant Ocean" (Aquarium Systems, Inc., Wickliffe, Ohio) has been the most widely used artificial seawater preparation. This mixture is adequate for the maintenance of a large number of species.

Artificial media offer the special advantage of allowing all experiments to be performed in a chemically defined environment. Knowledge of the chemical composition of the medium is a vital prerequisite for many analytical procedures. Even in those cases where the knowledge per se appears to be of little or no value, the use of a chemically defined medium eliminates many experimental variables.

In addition to such media as "Instant Ocean," which is a complex mixture, Fulton (1960) and Tusov and Davis (this volume) used simple artificial solutions containing only a few salts. For example, in "CCS-5" (the medium employed by Fulton to rear *Cordylophora*), Ca^{++}, Mg^{++}, Na^+, and K^+ are the only cations, and Cl^- and HCO_3^-, the only anions. The medium used by Tusov and Davis (this volume) to rear *Bougainvillia* contained Na^+, K^+, Mg^{++}, Mn^{++}, Ca^{++}, Cl^-, HCO_3^-, SO_4^{--}, and Br^-. Because of their simple composition, these media are extremely easy and inexpensive to use.

Preventing Stagnation

Two major methods have been developed to prevent stagnation in laboratory cultures of hydroid colonies. In one method, large volumes of solution are used and the medium is recirculated, filtered, and (usually) aerated continuously (see Crowell 1957). When this method is used, several colonies may be grown in the same container. The colonies usually are grown attached to 1" x 3" microscope slides held in glass slide racks. The racks, which are kept suspended in the recirculating system, can be conveniently transferred to small dishes containing *Artemia* nauplii whenever the hydroids are to be fed. After the animals feed, the slide racks are rinsed and returned to the recirculating system. The medium is changed at various intervals, depending upon the total volume of the system and the efficiency of the filtration device employed. Intervals which vary from 1 week (Crowell 1957) to 8 weeks (Hale 1964) between medium changes have been reported.

In the second method, individual colonies are grown in a relatively small volume of medium (a few hundred milliliters, at most). Because the medium is not circulated, it must be changed rather frequently (about twice each day). Less frequent changes of the medium are required if the polyps are not fed every day, or if the medium is agitated continuously— as with the rocker-arm assembly used by Russell and Rees (1936) or by placing the containers on a rotary shaker (Hauenschild and Kanellis 1953). Some species have been grown attached to the bottoms of the culture dishes (Nagao 1962; Kukinuma 1966a; Yamada 1961), but, in general, better results were obtained when the colonies were allowed to attach to some other surface, such as slanted microscope slides (Fulton 1960) or small glass rods (Hauenschild and Kanellis 1953; Müller 1964).

Regardless of the nature of the food, the composition of the medium, or the general culture system employed, colonial hydroids can be cultured successfully only if they receive daily care and attention.

GROWTH RATES

A typical hydroid colony consists of one or more vertical stems (hydrocauli), connected at their bases by a horizontal stolonal system, and having few to many lateral branches bearing hydranths (Figure 3). In some species (e.g., *Clytia johnstoni, Podocoryne carnea*), however, hydranths arise individually as vertical buds from the horizontal stolons (Figure 4).

Growth rates of colonial hydroids have been determined by measuring the rate at which (*a*) the total number of hydranths increases; (*b*) the individual hydranths form; (*c*) the length of individual stolons increases; and (*d*) the total length of all stolonal elements (stolons, stems, and lateral branches) in a colony increases.

Rate of Increase in Number of Hydranths

When measured as the increase in hydranth number, growth of colonial hydroids is exponential, and the logarithmic growth-rate constants (k) of most species fall in the range 0.20 to 0.35 (Table 2). These values indicate that the number of hydranths doubles every 2 to 3 days. However, if the colonies are allowed to continue growing over a longer period of time, or until they become extremely crowded, the growth rate eventually falls away from the exponential (Fulton 1962). The rate of hydranth production is known to be influenced by several factors; those producing the most obvious and immediate effects are temperature, salinity, nutritional level, and ionic composition of the medium (Fulton 1962; Crowell 1957; Crowell and Wyttenbach 1957; Tusov and Davis, this volume).

Rate of Hydranth Development

Crowell (1957) measured the time required for individual hydranths to develop in *Campanularia flexuosa* and found that this varied from 24 hours at 21° C to 43 hours at 10.5° C. According to Crowell, however, once a hydranth bud is initiated, the time required for its development into a mature hydranth is independent of variations in nutritional levels or age of upright. Therefore, these factors must influence growth rates by determining the rate at which new hydranth buds are initiated, rather than their rate of development. Whether this is true for other hydroids is not known.

Stolonal Growth Rates

The individual stolons of most species examined increase in length at rates ranging from 3 to 7 mm per day (Table 2). Considering the variety of

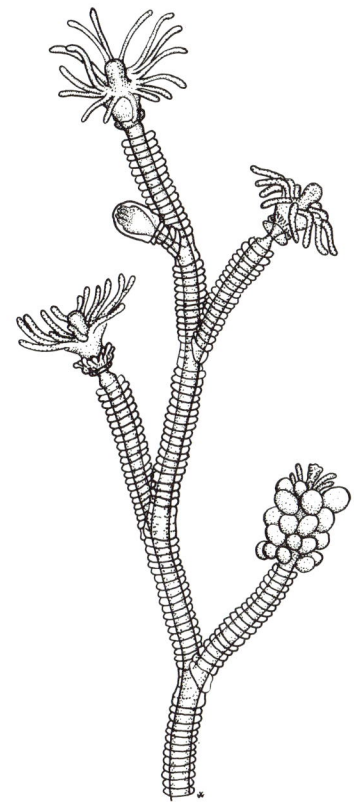

FIG. 3. Example of generalized upright growth form of colonial hydroids. Drawing by F.V. White.

conditions under which these observations were made, the similarity between these growth rates is remarkable, especially since stolonal growth rates vary with changes in the external environment. In *Obelia geniculata*, for example, free stolons (long, stolonlike outgrowths which are not attached to the substrate) grow at a rate of 1.0 mm/day at 10°–12° C, 10.0 mm/day at 16°–17° C, and 15–20 mm/day at 20° C (Berrill 1949a).

In *Cordylophora* (Fulton 1963a) and *Campanularia* (Wyttenbach 1968), newly initiated stolons grow exponentially for the first 4 to 6 days, after which the growth rate levels off to a constant linear rate. According to Kinne and Paffenhöfer (1966), however, individual stolons of *Clava multicornis* continue to grow exponentially for at least 40 days (the duration of their experiments), although with a logarithmic growth constant of only 0.045.

In all species, including those in which individual stolons grow linear rate, the growth rate of all of the stolonal elements in

FIG. 4. Example of "hydrolized" growth, typical of various species of colonial hydroids. Drawing by F.V. White.

considered together is exponential (Table 2). Fulton (1961) explained this situation, at least for *Cordylophora*, by showing that secondary stolons arise irregularly as lateral outgrowths from preexisting stolons, and that these secondary stolons, together with the stems and branches to which they give rise, appear frequently enough to make the growth of the entire colony exponential. Since the irregular production of lateral secondary stolons is common in most species of colonial hydroids, Fulton's explanation may have general significance.

Fulton (1961, 1963*a*) demonstrated also that the various stolonal elements of *Cordylophora* do not grow at the same rate. Thus, under the same conditions in which horizontal stolons grew at a rate of 3–4 mm/day, vertical stolons (= stems) grew only about half as fast (1.5 to 2 mm/day) and lateral branches grew only about one-sixth to one-eighth as fast (approximately 0.5 mm/day). In *Tubularia crocea*, on the other hand, both stolons and stems were observed to grow at a rate of 1.0 mm/day (Mackie 1966). Because the relative growth rates of different portions of a colony are related to the overall patterns of colony growth, it is unfortunate that similar data are not available for more species. A knowledge of intrinsic differences in the growth rates of the various elements in a colony might help to determine the manner in which colony form is influenced by environmental factors, and to distinguish between intrinsically and extrinsically regulated differences in growth patterns.

The effect produced by any environmental factor differs according to whether it influences each of the various stolonal elements to the same relative extent, or whether any single element is affected disproportionately. In the case of the former, the size of the colony at any given time is changed, but the overall pattern remains constant. In the case of the latter, however, changes in colony form can occur.

It has been suggested by Fulton (1963a) that a phenomenon of this nature might account for the alterations in colony form observed in *Cordylophora* in response to a change in salinity (Kinne 1958) and in *Campanularia* in response to variation in nutritional levels (Crowell 1957). The differences observed in the growth of *Bougainvillia* sp. in 50-percent and 100-percent seawater also might be explicable on this basis (Tusov and Davis, this volume). In this species, a decrease in salinity reduces the growth of stolons to a relatively greater extent than it does the growth of stems, so that colonies grown in 50-percent seawater form proportionately more stem than stolon, whereas in 100-percent seawater considerably more stolon than stem is formed. As a result, colonies grown in 50-percent seawater tend to be slightly more erect and less spread out than those grown in 100-percent seawater. A similar response was found in this species to temperature; that is, colonies grown at temperatures above 28° C tended to be somewhat lower and more spread out than those grown at temperatures below 28° C.

MORPHOGENETIC ACTIVITIES DURING STOLONAL ELONGATION*

Our present knowledge of the morphogenetic activities involved in stolonal elongation is extremely limited and fragmentary. Only a few studies have been made in this area, and even these few frequently contradict one another.

According to Berrill (1949a, b, c, 1950, 1952, 1953a, b), stolonal elongation in all species of colonial hydroids is the result of mitotic activity in both the ectodermal and endodermal cells located in a terminal meristematic region of the stolon. Chapman (1937) also reported that a terminal growth zone characterized by high levels of mitotic activity in both cell layers is present in *Obelia*. Müller (1964) reported similar findings for the early stages of stolon growth in *Hydractinia echinata*, although the terminal meristematic region in this species is said to consist of a mass of interstitial cells from which both ectodermal and endodermal cells are derived.

On the other hand, mitosis is reported to be absent from the terminal region of the stolons in *Clytia johnstoni* (Hale 1964), *Campanularia*

*In this section, the term stolonal growth (or stolonal elongation) will be used to describe the growth of all of the tubular stolonal elements in a colony, since the differences among these elements (horizontal stolons, stems, lateral branches, etc.) appear to be mostly positional.

(Crowell, Wyttenbach, and Suddith 1965; Wyttenbach 1965), *Cordylophora* (Overton 1963), and *Proboscidactyla* (Campbell 1968a, b).

Overton (1963) observed cells dividing in the region of the stolon in *Cordylophora* from which new vertical stems arise (approximately 3 mm behind the tip of the stolon). According to this author, this region is characterized by a dense accumulation of interstitial cells, and it was these cells, presumably, that were in the process of dividing.

Hale (1964) found several areas of localized mitotic activity in the stolons of *Clytia johnstoni*. These areas began at least 0.3 mm behind the stolon tip, and extended back for a distance of at least 15 mm. Mitosis was never seen in the first 0.3 mm of the stolon. In addition, the number of these areas, their position relative to each other or to either developing or developed polyps, and the number of dividing cells in each area were all extremely variable. In some stolons, no mitotic activity could be found. According to Hale, virtually all mitotic activity was confined to the ectodermal cells; only rarely were endodermal cells observed in the process of dividing, and no interstitial cells were found in this species. In *Proboscidactyla*, however, mitoses were observed in both ectodermal and endodermal cells (Campbell 1968b).

Thus, although it seems reasonably clear that stolonal growth does involve cell division, considerable confusion exists as to which cells are dividing and where those cells are located. With so few data available, it is impossible to determine whether the discrepancies that occur among these reports represent wrong interpretations on the part of some investigators or whether they indicate the existence of species-specific differences in the patterns of mitotic activity.

It is possible also that stolonal elongation involves the movement of cells into the advancing tip from elsewhere in the stolon. Hale (1964) determined, over a period of time, the position of marked ectodermal and endodermal cells in the stolons of *Clytia*, relative to a fixed point on the substrate and to the advancing stolon tip. By also determining the position of the stolon tip, he was able not only to detect the movement of ectodermal and endodermal cells, but also to determine the rate of this movement relative to the rate of stolonal elongation. He observed both cell layers in all growing stolons moving in the direction of the advancing tip. When the rate of cell movement was expressed as the ratio of cell movement to stolon elongation (rate of movement of the advancing tip), the following stolonal regions could be delineated:

1. A region located about 0 to 1 mm behind the stolon tip in which the marked cells moved toward the tip at a rate slightly faster than that of stolon elongation, so that ratios of 1.0 to 1.1 were obtained. The rate of movement of ectodermal cells on the tip of the stolon was slightly less than that for similar cells located on the sides or bottom of the stolon.

2. A region extending from approximately 1 to 3 mm behind the tip

in which the ectodermal cells moved considerably faster than the stolon tip itself, resulting in ratios ranging from 1.1 to 1.5. The difference in the rate of movement of the ectodermal cells located on the top of the stolon compared to those cells on the sides or bottom was more pronounced than the difference in the region 1 to 3 mm behind the tip to the tip, so that what had been a band of marked ectodermal cells became V-shaped, with the tip of the V on the top of the stolon and pointing away from the tip of the stolon. The endodermal cells in this region, although still moving in the same direction as the tip of the stolon, advanced more slowly than did the stolon, yielding ratios of 0.6 to 1.0.

3. A region located from about 2 or 3 mm behind the tip to a point some 4 to 6 mm behind the tip, in which measurements could not be made because the limits of the marked regions became indistinct as the marked cells spread out.

4. A region of variable length beginning about 4 to 6 mm behind the tip and extending to a point some 6 to 12 mm behind the tip. In this region both the ectodermal cells and the endodermal cells, although still moving in the same direction as the tip, moved more slowly than did the ectodermal and endodermal cells of the tip. Ratios of ectodermal cells in this region varied from 0.0 to 0.9, and those of the endodermal cells from 0.0 to 0.6, with the rates of movement in both layers decreasing with increasing distance from the tip until, eventually, the point was reached where no movement could be detected.

Movement of marked ectodermal cells has also been reported in *Cordylophora* stolons (Overton 1963). In this case, however, cellular movement was detected only in a region located some 1 or 2 mm behind the tip. Furthermore, the marked cells did not change their positions relative to the advancing stolon tip, which suggests that both the cells and the stolon tip were moving at the same rate. Other than these few observations, little else is known about the extent to which stolonal growth may depend upon cellular migration.

It has been suggested also that stolonal elongation, instead of being a morphogenetic phenomenon involving cellular proliferation or migration, may be due to the activities of contractile tissues located at or near the tip of the stolon (Hale 1960, 1964; Campbell 1968*a*). Although the details of how this mechanism might operate are not known, it may be hypothesized that two separate but coordinated activities are involved. The first of these would consist of a series of proximally directed waves of contraction that are initiated at the tip of the stolon and that operate against a solid substrate; the second would consist of rhythmic hydrostatic extensions of the stolon tip that arise from peristaltic contractions of the coenosarc in a region located about 0.5 to 1.5 mm behind the stolon tip, and which act against the fluid-filled hydrocoel. This latter activity has been known for some time and has been described in several different species of colonial hydroids (Berrill 1949a; Campbell 1966; Fulton 1963b; Hale 1960; Hudson 1965; Wyttenbach 1965, 1968).

The first process is known for certain to occur only in the holdfasts of *Corymorpha* (Campbell 1968a), although its existence in *Campanularia* is suggested by the observation that, when the tips of stolons are isolated distal to the main contractile region, they nevertheless retain the ability to creep along the substrate (Wyttenbach 1968). The small additional contractile region described by Hudson (1965) in the immediate vicinity of the stolon tips of *Clytia johnstoni* may also represent a similar mechanism.

The manner in which the stolon tips and the main contractile region interact is unknown, although Campbell (1968b) suggested that the terminal contractile region may act to translate the cyclic hydrostatic extensions into forward movement, and Wyttenbach (1968) postulated that the timing of the cyclic extensions is determined by the terminal region, the magnitude of each extension being a function of the subterminal contractile zone.

Whatever the mechanism is, it seems evident that stolonal growth is associated in some way with the rhythmic cycles of extension and contraction of the stolon tip; for stolonal growth, rather than being a smooth and continuous process, results from a series of extensions and retractions (Figure 5). In addition, Hale (1964) observed that stolonal growth in *Clytia* does not occur in the absence of activity in the major subterminal contractile zone.

It seems highly unlikely, however, that stolonal growth results entirely from the contractile properties of the tissues located within 1 or 2 mm of the tip. A more plausible explanation might be that stolonal growth is the product of three highly integrated processes: cellular proliferation, cellular migration, and mechanical elongation. If this is so, a delicate balance must exist among the rates of these three activities for stolonal growth to continue in a normal fashion. The division of cells and the migration of at least some of the products of division into the terminal region of the stolon would need to compensate exactly for the elongation produced by mechanical extension. If, for any reason, the mitotic and/or migratory activities were to cease or if they merely failed to keep pace with the mechanical elongation, the coenosarc in some regions behind the tip would become stretched. This stretching does occur in the holdfasts of *Corymorpha* which elongate without adding any new cells (Campbell 1968a). It was observed also that, as the holdfasts in *Corymorpha* continue to elongate, the tissues often stretch to the point where discontinuities appear. The older holdfasts are then frequently lost, presumably due to the severing of their connection to more proximal regions. Assuming that this elongation without the addition of new cells does occur in hydroid stolons, it is not difficult to visualize that frustules

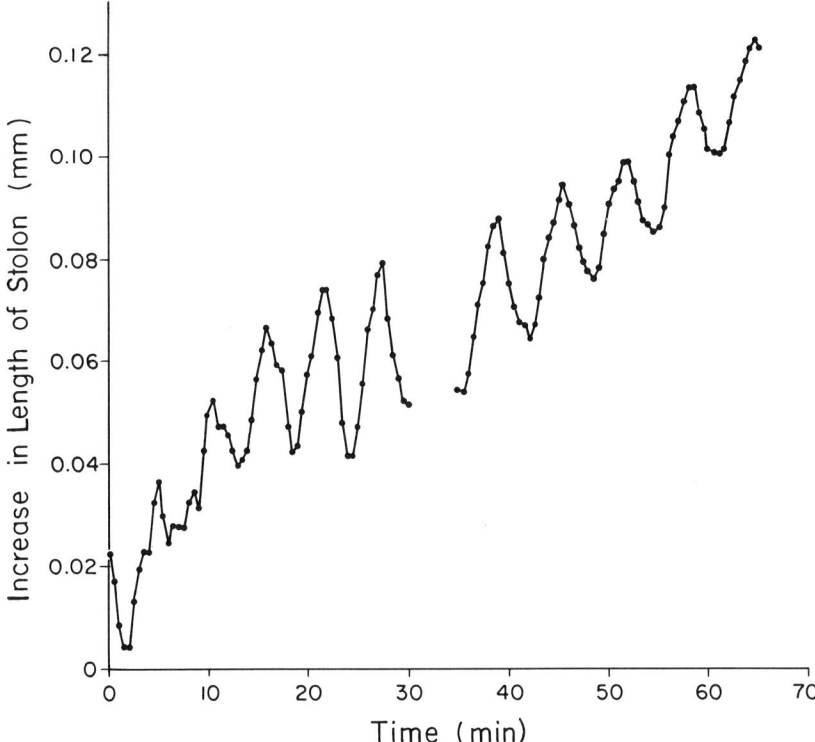

FIG. 5. Graph illustrating rhythmic cycles of extension and contraction of stolon tip in *Clytia johnstoni*. This is a characteristic form of stolonal elongation in colonial hydroids (Hale 1964). Reprinted by permission of the author.

may form in such genera as *Craspedacusta,* simply by developing in a manner similar to that of the holdfasts of *Corymorpha*.

If it is further assumed that the processes of cell division and migration are less influenced by temperature than is the rate of mechanical elongation (Wyttenbach 1968), this could account for the observations that at high temperatures stolonal growth appears to be favored over stem growth (Tusov and Davis, this volume). Under these conditions, more cells would be required to compensate for the greater rate of mechanical elongation. If mitosis and migration were not accelerated to the same extent, fewer cells would be available for stem formation, a process which also appears to depend to a large extent on the mitotic and migratory activities of cells in the stolon (Hale 1964).

At higher temperatures, it is conceivable that mitosis and migration would not proceed quickly enough to compensate for the increase in the rate of mechanical elongation. In this event cytoplasmic discontinuities would appear at some point behind the stolon tip, and the terminal portion of the stolon would eventually be lost. Although the underlying mechanism is unknown, this is, in fact, exactly what was observed to happen in the stolons of *Bougainvillia carolinensis* when colonies were grown at very high temperatures (Berrill 1949*b*).

However attractive this hypothesis might be, it is obvious that considerably more information is required before any definite conclusions can be drawn.

Aside from the fact that hydranth formation, like stolonal growth, appears to involve both cellular migration and cell division (Hale 1964; also reviewed by Tardent 1963), little is known of the various processes involved, the relative contributions of mitotic activity and cellular migration, or the actual mechanisms by which morphogenetic fields are established and differentiation of specific hydranth structures are controlled.

LITERATURE CITED

Berrill, N. J. 1949*a*. Polymorphic transformations in *Obelia*. *Quarterly J. Microscopical Science* **90**: 235-264.

———. 1949*b*. Growth and form in gymnoblastic hydroids. I. Polymorphic development in *Bougainvillia* and *Aselomaris*. *J. Morphology* **84**: 1-30.

———. 1949*c*. Growth and form in calyptoblastic hydroids. I. Comparison of a campanulid, *Campanularia, Sertularia,* and *Plumularia*. *J. Morphology* **85**: 297-335.

———. 1950. Growth and form in calyptoblastic hydroids. II. Polymorphism within the Campanularidae. *J. Morphology* **87**: 1-26.

———. 1952. Growth and form in gymnoblastic hydroids. II. Sexual and asexual reproduction in *Rathkea*. III. Hydranth and gonophore development in *Pennaria* and *Acaulis*. IV. Relative growth in *Eudendrium*. *J. Morphology* **90**: 1-32.

———. 1953*a*. Growth and form in gymnoblastic hydroids. VI. Polymorphism within the Hydractiniidae. *J. Morphology* **92**: 241-272.

———. 1953*b*. Growth and form in gymnoblastic hydroids. VII. Growth and reproduction in *Syncoryne* and *Coryne*. *J. Morphology* **92**: 273-302.

———. 1961. *Growth, development and pattern.* San Francisco: W. H. Freeman and Co.

Braverman, M. 1962. Studies on hydroid differentiation. I. *Podocoryne carnea* culture methods and carbon dioxide induced sexuality. *Experimental Cell Research* **26**: 301-306.

———. 1963. Studies on hydroid differentiation. II. Colony growth and initiation of sexuality. *J. Embryology and Experimental Morphology* **11**: 239-253.

Brinckmann, A. 1962. The life cycle of *Merga galleri* sp. n. *Pubblicazioni della Stazione zoologica di Napoli* **33**: 1-9.

———. 1964. Observations on the biology and development of *Staurocladia portmanni* sp. n. (Anthomedusae, Eleutheridae). *Canadian J. Zoology* **42**: 693-705.

Brinckmann-Voss, A. 1966. The morphology and development of *Acaulis ilonae* sp. n. (Anthomedusae/Athecata, Fam. Acaulidae). *Canadian J. Zoology* **44**: 291-301.

Brock, M. A., and B. L. Strehler. 1963. Studies on the comparative physiology of aging. IV. Age and mortality of some marine cnidaria in the laboratory. *J. Gerontology* **18**: 23-28.

Browne, E. T. 1897. On keeping medusae alive in an aquarium. *J. Marine Biological Association, United Kingdom* **5**: 176-180.

———. 1907. A new method for growing hydroids in small aquaria by means of a continuous current tube. *J. Marine Biological Association, United Kingdom* **8**: 37-43.

Campbell, R. D. 1966. Mechanisms of hydrozoan stolon elongation. *American Zoologist* **6**: 330.

———. 1968*a*. Holdfast movement in the hydroid *Corymorpha palma*: Mechanism of elongation. *Biological Bulletin, Woods Hole* **134**: 26-34.

———. 1968*b*. Colony growth and pattern in the two-tentacled hydroid, *Proboscidactyla flavicirrata*. *Biological Bulletin, Woods Hole* **135**: 96-104.

Chapman, S. S. 1937. Localization of -SH and -S.S- in *Obelia geniculata*. *Growth* **1**: 299-307.

Crowell, S. 1953. The regression-replacement cycle of hydranths of *Obelia* and *Campanularia*. *Physiological Zoölogy* **26**: 319-327.

———. 1957. Differential responses of growth zones to nutritive level, age, and temperature in the colonial hydroid *Campanularia*. *J. Experimental Zoology* **134**: 63-90.

———. 1961. Developmental problems in *Campanularia*. In *The biology of hydra and of some other coelenterates: 1961*, H. M. Lenhoff and W. F. Loomis, eds., pp. 297-316. Coral Gables: University of Miami Press.

Crowell, S., and C. Wyttenbach. 1957. Factors affecting terminal growth in the hydroid *Campanularia*. *Biological Bulletin, Woods Hole* **113**: 233-244.

Crowell, S., C. Wyttenbach, and R. L. Suddith. 1965. Evidence against the concept of growth zones in hydroids. *Biological Bulletin, Woods Hole* **129**: 403.

Fulton, C. 1960. Culture of a colonial hydroid under controlled conditions. *Science* **132**: 473-474.

———. 1961. The development of *Cordylophora*. In *The biology of hydra and of some other coelenterates: 1961*, H. M. Lenhoff and W. F. Loomis, eds., pp. 287-296. Coral Gables: University of Miami Press.

———. 1962. Environmental factors influencing the growth of *Cordylophora*. *J. Experimental Zoology* **151**: 61-78.

———. 1963*a*. The development of a hydroid colony. *Developmental Biology* **6**: 333-369.

———. 1963*b*. Rhythmic movements in *Cordylophora*. *J. Cellular and Comparative Physiology* **61**: 39-52.

Hale, L. J. 1960. Contractility and hydroplasmic movements in the hydroid *Clytia johnstoni*. *Quarterly J. Microscopical Science* **101**: 339-350.

———. 1964. Cell movements, cell division, and growth in the hydroid *Clytia johnstoni*. *J. Embryology and Experimental Morphology* **12**: 517-538.

Hancock, D. A., R. E. Drinnan, and W. N. Harris. 1956. Notes on the biology of *Sertularia argenta* L. *J. Marine Biological Association, United Kingdom* **35**: 307-325.

Hauenschild, C. 1954. Genetische und Entwicklungsphysiologische Untersuchungen über Intersexualität und Gewebeverträglichkeit bei *Hydractinia echinata* Flemm. (Hydrozoa; Bougainvill.). *Wilhelm Roux' Archiv für Entwicklungsmechanik der Organismen* **147**: 1–41.

Hauenschild, C., and A. Kanellis. 1953. Experimentelle Untersuchungen an Kulturan von *Hydractinia echinata* Flemm., zur Frage der Sexualität und Stockdifferenzierung. *Zoologische Jahrbücher*. Abteilung 3, *Allegemeine Zoologie und Physiologie* **64**: 1–13.

Hudson, R. C. L. 1965. A new contractile region in the stolon of the hydroid *Clytia johnstoni. Biological J.* **5**: 14–22.

Kinne, O. 1956. Über den Einfluss des Salzgehaltes und der Temperatur auf Wachstum, Form und Vermehrung bei dem Hydroidpolypen *Cordylophora caspia* (Pallas), Thecata, Clavidae. *Zoologische Jahrbücher*. Abteilung 3, *Allegemeine Zoologie und Physiologie Tiere* **67**: 407–486.

———. 1958. Über die Reaktion erbgleichen Coelenteratengewebes auf verschiedene Salzgehalts und Temperaturbedingungen. *Zoologische Jahrbücher*. Abteilung 3, *Allegemeine Zoologie und Physiologie* **67**: 407–486.

Kinne, O., and G. A. Paffenhöfer. 1965. Hydranth structure and digestion rate as a function of temperature and salinity in *Clava multicornis* (Cnidaria, Hydrozoa). *Helgoländer Wissenschaftliche Meeresuntersuchungen* **12**: 329–341.

———. 1966. Growth and reproduction as a function of temperature and salinity in *Clava multicornis* (Cnidaria, Hydrozoa). *Helgoländer Wissenschaftliche Meeresuntersuchungen* **13**: 62–72.

Kukinuma, Y. 1966*a*. Life cycle of a hydrozoan, *Sarsia tubulosa* (Sars) *Bulletin. Marine Biological Station of Asamushi, Japan* **12**: 207–210.

———. 1966*b*. Life cycle of a hydrozoan, *Campanulina* type, or *Aequorea coerulescens* Brandt. *Bulletin. Marine Biological Station of Asamushi, Japan* **12**: 211–218.

Loomis, W. F. 1953. The cultivation of hydra under controlled conditions. *Science* **117**: 565–566.

———. 1954. Environmental factors controlling growth in hydra. *J. Experimental Zoology* **126**: 223–234.

Lytle, C. F. 1961. Patterns of budding in the fresh water hydroid *Craspedacusta*. In *The biology of hydra and of some other coelenterates: 1961*, H. M. Lenhoff and W. F. Loomis, eds., pp. 317–336. Coral Gables: University of Miami Press.

Mackie, G. O. 1966. Growth of the hydroid *Tubularia* in culture. In *The Cnidaria and their evolution,* W. J. Rees, ed., pp. 397–410. New York: Academic Press.

Matthews, D. C. 1966. A comparative study of *Craspedacusta sowerbyi* and *Calpasoma dactyloptera* life cycles. *Pacific Science* **20**: 246–259.

McClary, A. 1959. The effect of temperature on growth and reproduction in *Craspedacusta sowerbyi. Ecology* **40**: 158–162.

Miller, R. 1966. Chemotaxis during fertilization in the hydroid *Campanularia. J. Experimental Zoology* **162**: 23–44.

Müller, W. 1961. Untersuchungen zur Stockdifferenzierung von *Hydractinia echinata. Zoologische Jahrbücher*. Abteilung 3, *Allegemeine Zoologie und Physiologie* **69**: 317–324.

———. 1964. Experimentelle Untersuchungen über Stockentwicklung, Polypendif-

ferenzierung und Sexualchimären bei *Hydractinia echinata. Wilhelm Roux' Archiv für Entwicklungsmechanik der Organismen* **155**: 181-268.
Muscatine, L., and H. M. Lenhoff. 1965. Symbiosis of hydra and algae. I. Effects of some environmental cations on growth of symbiotic and aposymbiotic hydra. *Biological Bulletin, Woods Hole* **128**: 415-424.
Muus, K. 1966. Notes on the biology of *Protohydra leuckarti* Greef (Hydroidea, Protohydridae). *Ophelia* **3**: 141-150.
Nagao, Z. 1962. The polyp and medusa of the hydrozoan *Stauridiosarsia japonica* n. sp., from Akkeshi, Hokkaido. *Annotationes zoologicae japonenses* **35**: 176-181.
———. 1964. The life cycle of the hydromedusa *Nemopsis dofleini* Maas, with a supplementary note on the life history of *Bougainvillia superciliaris* (L. Agassiz). *Annotationes zoologicae japonenses* **37**: 153-162.
Overton, J. 1963. Intercellular connections in the outgrowing stolon of *Cordylophora. J. Cell Biology* **16**: 661-671.
Palincsar, E. E., and J. S. Palincsar. 1960. The effect of 8-azaguanine and chloramphenicol on the regression-replacement cycle of hydranths. *Biological Bulletin, Woods Hole* **119**: 329-330.
Rees, W. J. 1936. On a new species of hydroid, *Staurocoryne filiformis*, with a revision of the genus *Staurocoryne* Rotch, 1872. *J. Marine Biological Association, United Kingdom* **21**: 134-142.
———. 1938. Observations on British and Norwegian hydroids and their medusae. *J. Marine Biological Association, United Kingdom* **23**: 1-42.
———. 1939. The hydroid of the medusa *Dipurena halterata* (Forbes). *J. Marine Biological Association, United Kingdom* **23**: 343-346.
———. 1941. Notes on British and Norwegian hydroids and medusae. *J. Marine Biological Association, United Kingdom* **25**: 129-141.
Rees, W. J., and F. S. Russell. 1937. On rearing the hydroids of certain medusae, with an account of the methods used. *J. Marine Biological Association, United Kingdom* **22**: 61-82.
Reisinger, E. 1957. Zur Entwicklungsgeschichte und Entwicklungsmechanik von *Craspedacusta* (Hydrozoa; Limnotrachylina). *Zeitschrift für Morphologie und Ökologie der Tiere* **45**: 656-698.
Roch, F. 1924. Experimentelle Untersuchungen von *Cordylophora caspia* (Pallas) (= *lacustris* Allman) über die Abhängigkeit ihrer geographischen Verbreitung und ihrer Wuchsformen von der physikalischchemischen Bedingungen des umgebendes Mediums. *Zeitschrift für Morphologie und Ökologie der Tiere* **2**: 350-426.
Russell, F. S. 1936. On the first stage of the medusa *Echeilota clausa* (Hincks) (= *E. hartlaubi* Russell). *J. Marine Biological Association, United Kingdom* **21**: 131-133.
Russell, F. S., and W. J. Rees. 1936. On rearing the hydroid *Zanclea implexa* (Alder) and its medusa *Zanclea gemmosa* McCrady, with a review of the genus *Zanclea. J. Marine Biological Association, United Kingdom* **21**: 107-130.
Strehler, B. L. 1961. Aging in coelenterates. In *The biology of hydra and of some other coelenterates: 1961*, H. M. Lenhoff and W. F. Loomis. eds., pp. 373-398. Coral Gables: University of Miami Press.

Tardent, P. 1963. Regeneration in the hydrozoa. *Biological Reviews and Biological Proceedings of the Cambridge Philosophical Society* **38**: 293-333.
———. 1965. Ecological aspects of the morphodynamics of some hydrozoa. *American Zoologist* **5**: 525-529.
Toth, S. E. 1965. Cultivation of marine hydroids. *Bios* **36**: 63-65.
Weiler-Stolt, B. 1960. Über die Bedeutung der Interstitiellen Zellen für die Entwicklung und Fortpflanzung Mariner Hydroiden. *Wilhelm Roux' Archiv für Entwicklungsmechanik der Organismen* **152**: 398-454.
Wyttenbach, C. R. 1965. Sites of mitotic activity in the colonial hydroid, *Campanularia flexuosa*. *Anatomical Record* **151**: 483.
———. 1968. The dynamics of stolon elongation in the hydroid *Campanularia flexuosa*. *J. Experimental Zoology* **167**: 333-352.
Yamada, M. 1961. Polyp and medusa of *Podocoryne hartlaubi* Neppi and Stiasny (Hydrozoa) from the Gulf of Naples. *Pubblicazioni della Stazione zoologica di Napoli* **32**: 134-143.

S. ARTHUR REED CHAPTER 4
Michigan State University, East Lansing

Some Common Coelenterates in Kaneohe Bay, Oahu, Hawaii

The Hawaii Institute of Marine Biology (HIMB) is located on Coconut Island in Kaneohe Bay on the windward (eastern) side of the island of Oahu. The bay is roughly rectangular in shape being some 8 miles long and approximately 1 mile at its greatest width. It is bounded on the seaward side by a heavily eroded coral reef which, according to some geologists, represents the only true barrier reef in the Hawaiian Islands. A channel extending the length of the bay has been dredged to a depth of 40 feet, making it navigable to rather large ocean-going vessels. Two other shallower channels permit the passage of smaller boats into the open ocean. Erosion of the seaward coral reef has produced an extensive coral-rubble and sand-flat barrier where the water is only a few feet deep. This barrier effectively protects the bay from large ocean waves and tsunamis. Within the bay, a luxuriant growth of fringing reefs and patch reefs has developed. A few of these patch reefs are less than 100 feet in diameter and extend almost vertically from the bottom to a depth of 40 feet. Located on these reefs, and readily available to the research biologist, are a number of species of corals and other coelenterates that can serve as excellent research organisms.

The following list is a selection of the more common species, and is not intended to be exhaustive or all-inclusive. Comments are included on identifying characteristics, interesting points of natural history, and information on specific collecting sites in Kaneohe Bay of those species that are not generally distributed or easily located on the reefs.

ORDER SCLERACTINIA (STONY CORALS)

Over 120 species and varieties of corals have been collected and named from the waters around the Hawaiian Islands. Most of these, however, are deep-water forms obtained through dredging operations. Only a few species are common enough to be considered useful for research purposes.

Family Fungiidae

Three genera of this family of mushroom corals are common in Kaneohe Bay. One is extremely abundant and the other two are much less so, having been collected only at greater depths. These corals are solitary forms and are not attached in the adult stage to a basal stalk. They reproduce either sexually, producing a planula larval stage; or asexually, by budding from a parent polyp. The young polyp grows vertically from the base for a short period of time, forming an upright stalk. Later in development, the upper peripheral regions of the stalk grow laterally into a mushroom-shaped disc. As this disc enlarges, it breaks free from the basal stalk, and the organism takes on a free-living, unattached existence. The stump of the stalk often continues to produce more individuals, each of which breaks free after a period of growth.

Fungia scutaria. This species (Figures 1 and 2) is abundant on the reefs in Kaneohe Bay and large clusters of free-living individuals can be seen lying in crevices and pockets of the reef. The adult skeleton is oval in shape and can measure 6 to 7 inches in length. The young attached polyps are often lavender in color and change to deep brown as adults. The brown color may be due to the presence of large numbers of symbiotic algae, zooxanthellae, living in the endodermal tissues of the organism. In many specimens, the tentacles are deep green. The feeding behavior of these corals can be easily demonstrated by placing a small quantity of fresh or frozen brine shrimp on the tentacular surface of the animals. *Fungia scutaria* is hardy and will exist in an aquarium for several months with minimum care and feeding.

Fungia patella. The skeleton of this species (Figure 3) is nearly circular and the septa are slightly arched on the oral surface. The greatest diameter of most specimens does not exceed 2 inches. This coral can be collected only at depths of 40 to 60 feet on the gradually sloping, seaward side of the barrier reef running the length of Kaneohe Bay. Individuals brought up from this depth are usually a very light brown in color, but, when exposed to brighter light in an aquarium, rapidly darken to a deep brown, possibly because of an increase in zooxanthellae in the tissues. This species also does quite well in an aquarium.

Fungia fragilis. This species (Figure 4) has also been collected only at depths of 60 to 80 feet on the outer slope of the reef. A typical individual rarely exceeds 1 inch in diameter, and is characteristically deeply incised on one side rather than being circular. This appearance is apparently due to a fracturing of the skeleton along the radial septal lines. Many specimens collected exhibited a partial regrowth of the skeletal material.

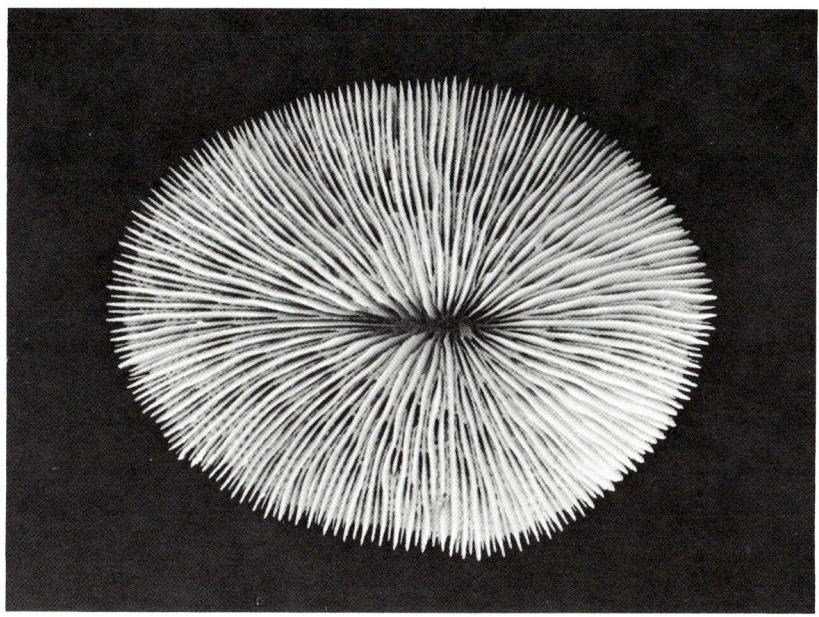

FIG. 1. *Fungia scutaria.* Skeleton. 3 inches.

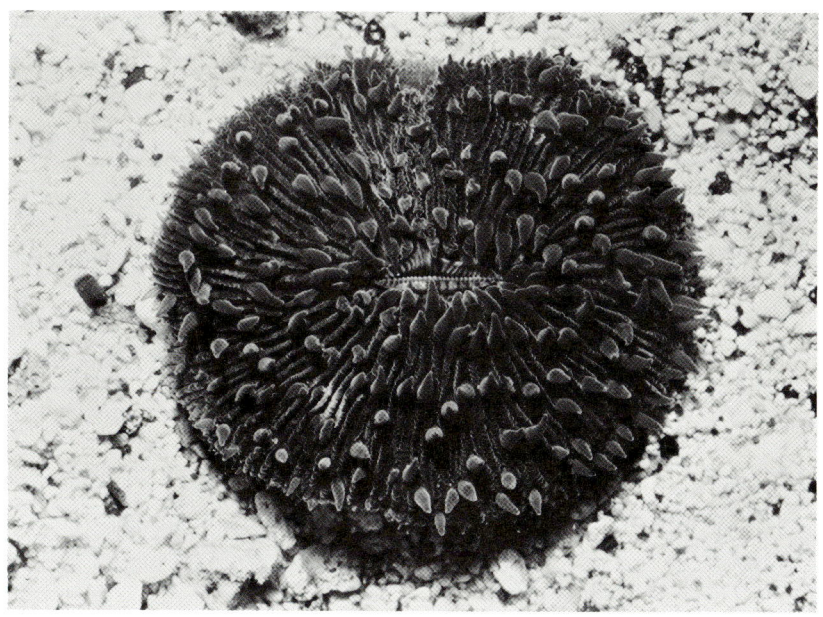

FIG. 2. *Fungia scutaria.* Live specimen. 3 inches.

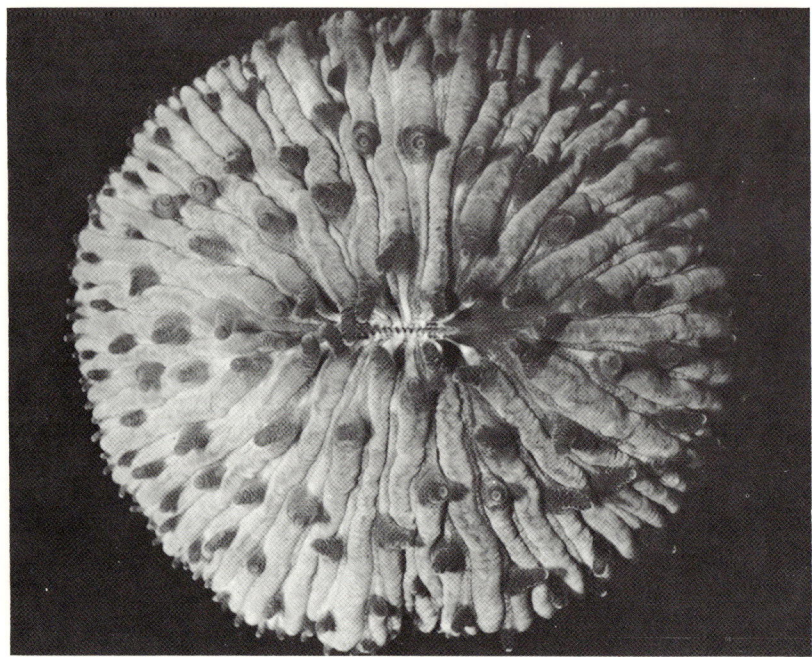

FIG. 3. *Fungia patella.* Live specimen. 1½ inches.

FIG. 4. *Fungia fragilis.* Live specimens. Size range from ½ to 1 inch.

Fungia (*hexagonalis?*). A few specimens have been collected which exhibit an unusual hexagonal-shaped skeleton. These have tentatively been identified as *Fungia hexagonalis* (Figure 5).

Family Poritidae

The bulk of most coral reefs in Kaneohe Bay consists of various species of corals in this family. Massive heads typically 10 to 15 feet across are characteristic, especially in the northern part of the bay.

Identification of species within this group is difficult because of the considerable amount of variation in growth forms. Microscopic examination of details of structural features of the skeleton is usually necessary. The two most common species, making up the greatest portion of the reef structure, are *Porites compressa* and *Porites lobata*.

Individual polyps of these colonial corals are quite small and produce copious amounts of slime on their outer surface, especially when disturbed. Both of those characteristics make it difficult to work with the tissues.

Some interesting cystlike structures which develop at the basal region of individual polyps in the species *Porites compressa* have recently been

FIG. 5. *Fungia hexagonalis* (?). Live specimen with tentacles contracted. 1 inch.

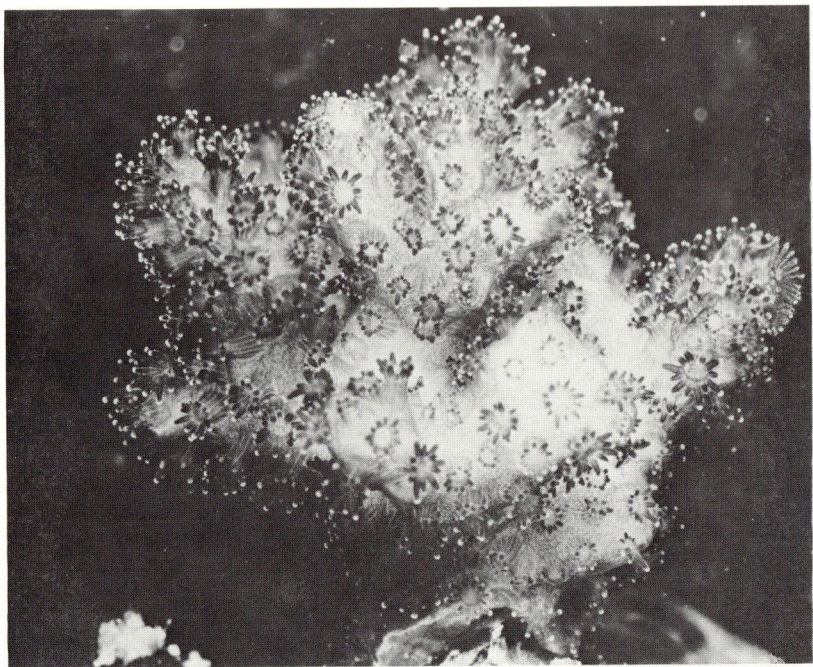

FIG. 6. *Pocillopora damicornis.* Live colony with polyps extended. 1½ inches.

FIG. 7. *Pocillopora damicornis.* Skeleton of delicate growth form. 4 inches.

discovered (A. Szmant and T. Cheng, unpublished). These small round objects, possibly a nematode in the encysted stage, caused a pink discoloration of the living polyp and caused the coral to secrete a highly modified skeletal structure. Infected specimens can be collected routinely on the reef adjacent to buoy 8 which marks the southern entrance, sometimes called Sampan Channel, to the bay.

Family Seriatoporidae

Two common species in this family are *Pocillopora damicornis* (Figures 6 and 7), and *Pocillopora meandrina* (Figure 8). *Pocillopora damicornis* offers many desirable features for research purposes. The fingerlike branches of the corallum are long and delicate, with individual polyps of the colony rather widely separated from each other. Little slime is produced. Small chunks of the coral head can be broken off and easily maintained in an aquarium for extended periods. The polyps readily extend their tentacles and actively feed on a variety of food in the laboratory. During the summer months, planula larvae can be collected in large numbers by removing coral heads from the reef and placing them in aquaria. *P. damicornis* grows abundantly on most of the reefs in Kaneohe Bay, and heads of varying diameters are available. An especially delicate growth of this species is common on the patch reef near buoy 14 in the north end of the bay.

Pocillopora meandrina. This species, which produces a more robust skeleton, grows on the outer slopes of the barrier reef in Kaneohe Bay. An interesting feature of this coral is the formation of inflated gall-like enlargements of the skeleton which contain a small female crab of the

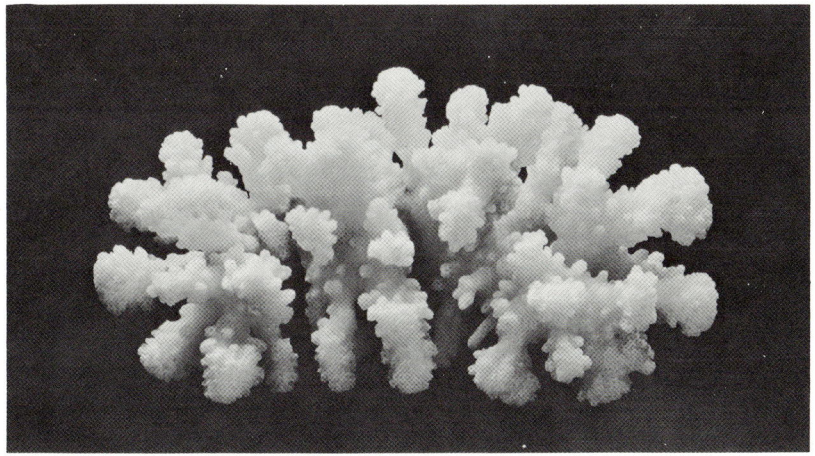

FIG. 8. *Pocillopora meandrina.* Skeleton. 6 inches.

FIG. 9. *Tubastrea aurea* (*Dendrophyllia manni*). Skeleton. 2½ inches.

FIG. 10. *Tubastrea aurea* (*Dendrophyllia manni*). Live colony with polyps extended. 2 inches.

species *Hapalocarcinus marsupialis*. The coral grows until it completely imprisons the crab, leaving only a few small openings through which food and water can enter.

Family Dendrophyllidae

The one species of this family common to Hawaiian waters is *Tubastrea aurea* (Figures 9 and 10), formerly called *Dendrophyllia manni*. This coral is brightly colored, ranging from bright orange to red, pink, and a greenish brown.

There are no zooxanthellae in the tissues of this ahermatypic coral. It does not contribute significantly to the reef formation. Small colonies of this species can be found growing under ledges on many of the patch reefs in Kaneohe Bay, especially on the reefs near buoys 14 and 9 near the main channel. They seem to thrive best in diffused light. The coral grows well in an aquarium, feeding most readily at night. Polyps can often be made to expand in daylight by placing a small amount of brine shrimp on the contracted organism. Planula larvae have been collected from *Tubastrea*, but in smaller numbers than have been collected from *Pocillopora damicornis*. The larvae, when expelled, are the same color as the parent polyp.

ORDER ACTINIARIA (SEA ANEMONES)

A few species of sea anemones are common on the reefs and can be collected in great number.

Aiptasia sp. This species grows abundantly in the shallower parts of the reef face, on the reef flat on Coconut Island, and throughout Kaneohe Bay (Figure 11). Living specimens are hardy and can easily be maintained in aquaria with minimum care for long periods. They readily capture and ingest frozen or live brine shrimp. This anemone reproduces asexually by pedal laceration: the outer edge of the pedal disc firmly attaches to the substratum, the central portion then pulls away, leaving behind small bits of the pedal disc. Each piece of tissue develops into another polyp.

Macranthea cookei. This is one of the largest anemones in the bay, sometimes averaging 10–30 cm from base to oral disc (Figure 12). The organism characteristically extends its column into the coral sand on the reef flats, exposing only the tentacle-covered oral disc. When disturbed, the oral disc closes and the column retracts, causing the anemone virtually to disappear into the sand.

Boloceroides lilae. This anemone is much less common but numbers can be collected with diligent effort. The species has an ability, unusual in

FIG. 11. *Aiptasia* sp. Live polyp. 2 inches.

sea anemones, to detach its base rapidly and voluntarily from the substratum and to swim actively to another location. The swimming movement is a coordinated pulsation of the tentacular crown. The tentacles are often easily shed by constriction of a sphincter at their base. Some claim that a new individual can regenerate from the severed end of the detached tentacle.

ORDER ZOANTHIDEA

Two genera are common in Hawaii, *Palythoa* and *Zoanthus* (Figure 13), both exhibiting colonial growth. Some species are quite common, covering large areas of the reef in a continuous mat, whereas other species are rare and difficult to find. Members of these groups are toxic to humans in varying degrees, causing mild to severe skin irritation following contact.

FIG. 12. *Macranthea cookei.* Live polyp. 6 inches.

FIG. 13. *Zoanthus* sp. Live colony. Diameter of oral surface is ½ inch.

TAXONOMIC CHECKLIST OF SOME COMMON COELENTERATES IN KANEOHE BAY, OAHU

Phylum Cnidaria (Coelenterata)
    Class Hydrozoa
        Order Hydroida
            Family Pennariidae
                Pennaria tiarella. Common on floats and piers

Order Siphonophora
    Physalia utriculus. Occasionally blown in from open ocean
    Velella pacifica. Occasionally blown in from open ocean
Class Scyphozoa
    Order Rhizostomaceae
        Mastigias papua (Figure 14). Jellyfish periodically abundant in Kaneohe Bay
Class Anthozoa
    Subclass Zoantharia
        Order Actiniaria (sea anemones)
            Macranthea cookei. Common on sand flats
            Boloceroides lilae. Seasonally common on reef slopes
            Aiptasia sp. Common on reef slopes
        Order Madreporaria (Scleractinia) (stony corals)
            Family Faviidae
                Leptastrea purpurea. Rare in bay
                Leptastrea bottae. Fairly common in north end of bay
                Cyphastrea ocellina. Fairly common in north end of bay and buoy 8, Sampan Channel
            Family Seriatoporidae
                Pocillopora damicornis. Common on most reefs
                Pocillopora meandrina. Common on seaward side of barrier reef slope
            Family Acroporidae
                Montipora verrucosa. Very common on most reefs
                Montipora flabellata. Rare, outer slopes of barrier reef
            Family Fungiidae
                Fungia scutaria. Very common on most reefs
                Fungia patella. Rare, outer slopes of barrier reef
                Fungia fragilis. Rare, outer slopes of barrier reef
            Family Poritidae
                Porites compressa. Very common; makes up bulk of reef
                Porites lobata. Common in northern end of bay
            Family Agariciidae
                Pavona varians. Common on reef slope at a depth of 10 feet or more
                Pavona explanulata. Rare in bay. Present in Hanauma Bay
            Family Thamnasteriidae
                Psammacora stellata. Rare, some on reef flats
            Family Dendrophyllidae
                Tubastrea aurea (*Dendrophyllia manni*). Common on a few reefs in north end of bay
        Order Zoanthidea
            Zoanthus sp. Common on reef flat around Coconut Island
            Palythoa sp. Rare in bay
        Order Ceriantharia
            Cerianthus sp. Rare in bay

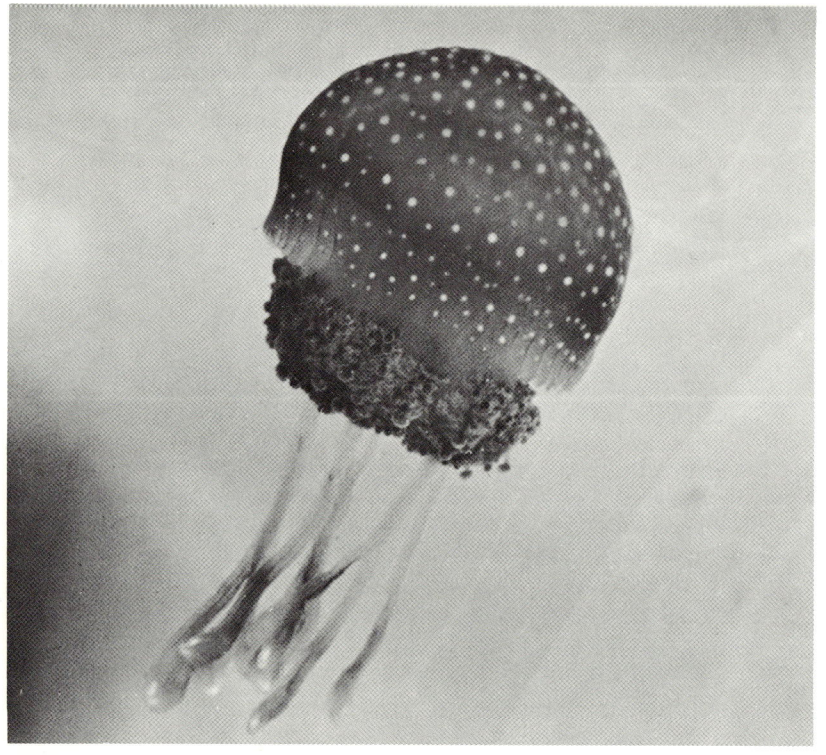

FIG. 14. *Mastigias papua.* Live medusa. Diameter of umbrella is 2½ inches.

SELECTED REFERENCES ON THE TAXONOMY, NATURAL HISTORY, AND PHYSIOLOGY OF COELENTERATES

Edmondson, C. H. 1933. *Reef and shore fauna of Hawaii.* Bernice P. Bishop Museum, special publication 22. Honolulu. (Two-volume revision in preparation.)

Florkin, M., and B. Scheer. 1968. *Chemical zoology.* Vol. 2, *Porifera, Coelenterata, and Platyhelminthes.* New York: Academic Press.

Hiatt, R. W. *Hawaiian marine invertebrates, a guide to their identification.* Unpublished key.

Hyman, L. H. 1940. *The invertebrates.* Vol. 1, *Protozoa through Ctenophora.* New York: McGraw Hill.

Lenhoff, H. M., and W. F. Loomis. 1961. *The biology of hydra and of some other coelenterates: 1961.* Coral Gables: University of Miami Press.

Moore, R. C., ed. 1956. *Treatise on invertebrate paleontology.* Vol. F, *Coelenterata.* Geological Society of America. Lawrence: University Press of Kansas.

Rees, W. J., ed. 1966. *The Cnidaria and their evolution*. New York: Academic Press.
Russell-Hunter, W. D. 1968. *A biology of lower invertebrates*. New York: MacMillan Co.
Vaughan, T. W., and J. W. Wells. 1943. Revision of the suborders, families, and genera of the Scleractinia. Geological Society of America, special paper no. 44. Lawrence: University Press of Kansas.

JOANNE TUSOV CHAPTER 5
University of California at Berkeley
LARY V. DAVIS
*Commission on Undergraduate Education in the Biological Sciences,
Washington, D.C.*

Influence of Environmental Factors on the Growth of *Bougainvillia* sp.

Most studies on colonial hydroids do not take into account the way in which the growth and development of these organisms are influenced by environmental factors. The absence of adequate information on growth under defined conditions has, on occasion, resulted in situations in which only highly qualified conclusions could be drawn from the vast amounts of data obtained (see, for example, Hammett 1950).

In order to avoid such a situation, the researcher must precede his developmental studies by observing, under defined conditions, the specific effects produced by alterations in various environmental factors. The present study was undertaken in an effort to provide such information for an unidentified species of the colonial hydroid *Bougainvillia*.

MATERIALS AND METHODS

A clone of *Bougainvillia* sp. was isolated from material collected in Kaneohe Bay, Oahu. Colonies of this species are composed of simple stems with irregular branching. A smooth, wavy perisarc covers the stolons and stems and extends up to the tentacular whorl, forming a pseudohydrotheca. The hydranths are conical with a single irregular row of 13 to 17 filiform tentacles. Gonophores are arranged singly on pedicels on the hydranth's stem. At the time of liberation the medusae have four manubrial tentacles which are beginning to branch. Located around the margin of the bell are four circular marginal bulbs, each of which gives rise to two simple tentacles. The marginal bulbs are brownish orange, and an ocellus is present at the base of each tentacle.

A few colonies developed gonophores and produced medusae when they were not fed regularly and the culture medium was not changed frequently. No colonies developed gonophores when maintained under standard conditions.

Subcultures were made by tying a single upright (stem and hydranth) to a microscope slide with a piece of thread (Crowell and Wyttenbach 1957). Within 2 days the stolonic outgrowth attached to the slide and the thread was removed. New colonies were then allowed to grow for 1 week before being used in an experiment, at which time they usually had 15 or more hydranths.

Bougainvillia colonies grow by developing new stolons and by budding new hydranths. The hydranths are spaced more or less evenly, and increase in stem length with age. On older uprights, lateral branching produces new hydranths and free stolons. When the free stolons touch the slide, they attach, grow, and give rise to more uprights. The hydranths do not regress in a healthy colony, and can be used as a quantitative estimate of growth. This method was used previously with *Hydra* (Loomis 1954), *Cordylophora* (Fulton 1962), and *Podocoryne* (Braverman 1963).

To reduce variability in growth rates, the following defined culture conditions, similar to those described by Fulton (1960), were followed throughout the experiments: all cultures were grown on microscope slides slanted in 100-ml beakers filled with either filtered seawater or an artificial salt solution (see Table 1). All cultures were maintained in the dark at 22° ± 1° C, and were fed daily with newly hatched brine shrimp nauplii. The culture medium was changed twice daily, once approximately 30 minues to 1 hour after feeding, and again approximately 6 hours later.

The culture medium was prepared by slightly modifying a formula for artificial seawater devised by Brujewicz (Subow 1931), and contained only the major elements found in seawater. The sulfate-ion source was as the sodium salt instead of the magnesium salt. The solution was prepared in demineralized water (Barnstead two-stage demineralizer, equipped with red-cap and black-cap cartridges) and then bubbled with air. The pH of the final solution was 7.5 to 8.1. The chlorinity of this solution was 19 parts per thousand.

When variations in either sulfate- , potassium- , magnesium- , or calcium-ion concentrations were tested, the medium was prepared without the ion in question. Known concentrations of the ion to be tested were then added to the solution. To test the effects of chloride ions, a special base medium containing Na_2SO_4, $Ca(NO_3)_2$, $MgSO_4$, $KHCO_3$, and NaBr was prepared. To this was added the appropriate concentration of NaCl.

Logarithmic growth rates were determined by Loomis's method

TABLE 1   Formula for artificial seawater

Salt	Grams/Liter	Moles/Liter
NaCl	25.6245	0.4384
$MgCl_2$	3.8092	0.04
Na_2SO_4	4.0061	0.0282
$CaCl_2$ (anhydrous)	1.1321	0.0102
KCl	0.7231	0.0097
$NaHCO_3$	0.2016	0.0024
NaBr	0.08295	0.000806

(1954). The number of hydranths in each colony, counted at the same time each day for 5 or more successive days, was plotted on semilog paper, and the best fitting straight line determined by regression analysis (Figure 1). The time required for the hydranth number to double was determined from the graph to the nearest tenth of a day. The logarithmic growth rate constant, k, was obtained using the formula $0.693/T = k$, where T is the doubling time in days.

Colonies subjected to unfavorable conditions resorbed their hydranths within a short period of time. This lack of growth was indicated as

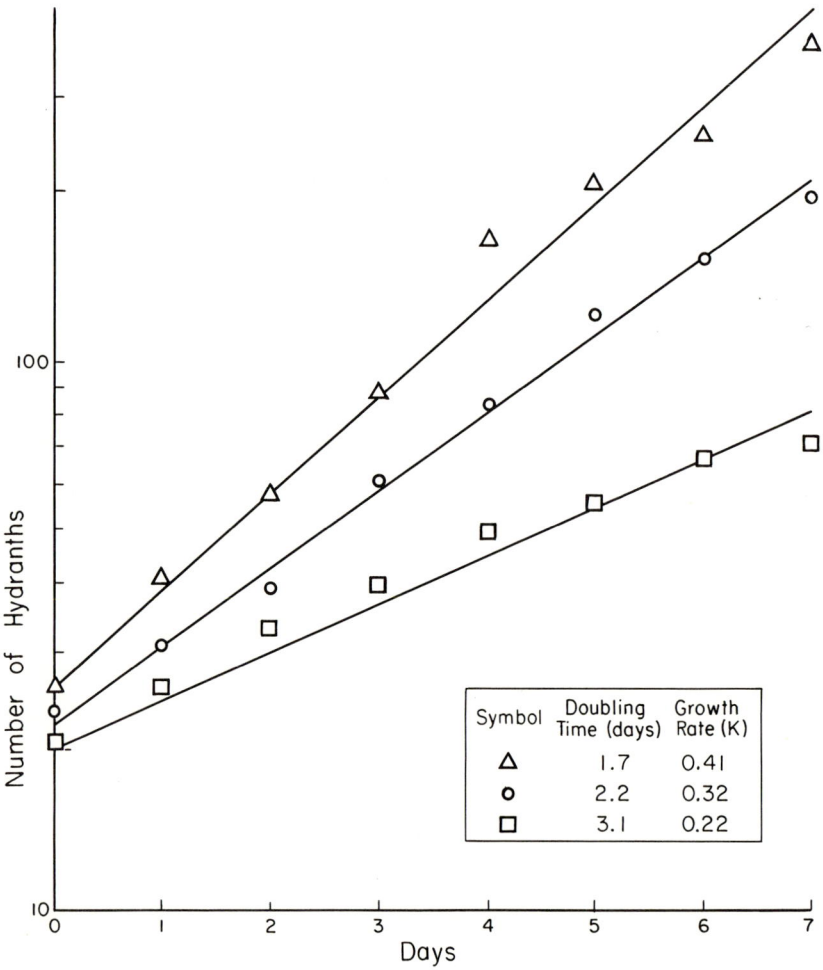

FIG. 1. Semilog plots of typical growth in *Bougainvillia* sp.

$k = 0.00$. When growth proceeded at a rate too low to measure, it was recorded as $k < 0.1$.

RESULTS

Growth of Bougainvillia under Standard Conditions

The mean growth rate obtained for all cultures grown under standard conditions was $k = 0.32$, which represents a doubling time of about 2 days. However, the growth of colonies under standard conditions sometimes varied considerably. To determine the amount of this variability, data from 10 experiments (43 cultures) were evaluated statistically (Table 2). The standard deviation from the mean growth rate (0.32) for these cultures was ±0.049. The mean growth rate of each group considered separately varied from 0.24 to 0.38, with a range from 0.03 to 0.13. The standard deviation for each group, estimated by calculating from the range, had a mean of ±0.034, which is slightly less than the standard deviation of the 43 cultures as a whole. This indicates that variability within a group is less than that among groups.

To estimate variability within an experiment, the researcher can calculate the range of growth to determine the significance of the k values among replicate cultures. Of the 10 groups, six had a range of 0.06 or less, and nine had a range of 0.09 or less. The mean range is 0.065 with a standard deviation of ±0.027. Therefore, 95 percent of the time a deviation greater than 0.09 from the mean rate of growth of the cultures grown under standard conditions can be considered a significant variation.

Growth in Filtered Seawater

Fifteen cultures from four experiments grown in filtered seawater had a mean growth rate of 0.31 and a range of ±0.08 (Table 3). These results show that growth rates obtained with an artificial culture medium, where the mean growth rate was 0.32 with a range of ±0.065 (Table 2), compare favorably with those obtained with natural seawater.

Salinity

Three experiments were conducted in which the relative concentration of artificial seawater was varied from 100 percent to 10 percent for replicate cultures maintained under standard conditions (Table 4). The mean growth rates of *Bougainvillia* in concentrations of seawater between 55 percent and 100 percent ranged from 0.27 to 0.41. These values are within the range of variation obtained for all cultures grown under standard conditions. At a seawater concentration of 50 percent, the growth rate was 0.25, which is slightly outside the normal range of variation, and, at 40 percent seawater, the growth rate declined to 0.13.

TABLE 2 Growth rates of replicate cultures

Group No.	No. of Cultures	Growth Rates (k)	Mean Growth Rate (k)	Range	Estimated σ
1	4	0.32, 0.32, 0.34, 0.27	0.31	0.07	0.034
2	10	0.30, 0.30, 0.32, 0.25, 0.25, 0.25, 0.25, 0.32, 0.26, 0.32	0.28	0.07	0.023
3	3	0.33, 0.39, 0.36	0.36	0.06	0.035
4	7	0.29, 0.33, 0.41, 0.28, 0.34, 0.36, 0.36	0.34	0.13	0.048
5	2	0.35, 0.41	0.38	0.06	0.053
6	4	0.35, 0.35, 0.39, 0.36	0.36	0.04	0.019
7	4	0.27, 0.22, 0.25, 0.25	0.24	0.05	0.024
8	3	0.39, 0.39, 0.30	0.36	0.08	0.047
9	3	0.32, 0.35, 0.35	0.34	0.03	0.018
10	3	0.32, 0.26, 0.32	0.30	0.06	0.035
Mean Values			0.32	0.065	0.034

NOTE: This table includes data for all cultures grown in artificial seawater under standard conditions over a period of 2 months.

TABLE 3 Growth rates of *Bougainvillia* in filtered seawater

Group No.	No. of Replicate Cultures	Growth Rates (k)	Mean Growth Rate (k)	Range
1	5	0.26, 0.30, 0.27, 0.25, 0.32	0.28	0.05
2	3	0.36, 0.43, 0.36	0.38	0.07
3	4	0.26, 0.36, 0.22, 0.32	0.29	0.10
4	3	0.33, 0.25, 0.27	0.28	0.08
Mean Values			0.31	0.08

The colonies did not survive for more than 24 hours at lower concentrations. The low growth rate in 40 percent seawater suggests that either certain ions become limiting at this concentration or that the osmotic pressure was too low.

Effect of Salinity on Relative Lengths of Stems and Stolons

To determine whether the number of uprights and hydranths also reflected the amount of horizontal stolonic growth at the various salinities, colonies grown in 50 percent, 70 percent, and 100 percent artificial seawater were photographed at the beginning of the experiment and then once each day for 3 days. From the photographs, tracings of stolons and stems were measured to obtain their total lengths on each day (Table 5). Colonies grown in 100 percent and in 70 percent seawater produced new stems and stolons in approximately the same relative lengths as were present initially, so that there was little change in relative lengths during the 3 days of growth. Similarly, the length of stolon per hydranth present in these colonies remained relatively constant at both concentrations. After 3 days' growth in 50 percent artificial seawater, however, the relative lengths of stolon and stem decreased and increased, respectively; also, the length of stolon per hydranth decreased. The effects of diluting the seawater to 50 percent are evident also in a comparison of the new growth that occurred during the experiment (Table 5). Of the new tubular growth, only 48.5 percent went into the formation of new stolons, while 51.5 percent appeared as stem.

Ionic Requirements

The data in Figure 2 show that *Bougainvillia* has an absolute requirement for potassium, calcium, magnesium, and chloride ions. (All of the experimental solutions had a final molarity of between $0.45\ M$ and $0.556\ M$.) The majority of colonies that did not grow regressed within 1 or 2 days.

In colonies grown in low concentrations of potassium, the cells in

TABLE 4 Growth rates by dilution of seawater

Relative Seawater Concentration		No. of Cultures	Mean Growth Rate (k)	Range	Variation from Mean k of Controls
%	Moles/Liter				
100	0.504	3	0.36	0.06	—
100	0.497	6	0.34	0.12	—
100	0.493	2	0.38	0.06	—
90	0.457	3	0.40	0.02	0.04
85	0.445	3	0.40	0.01	0.04
80	0.410	3	0.41	0.00	0.05
80	0.405	2	0.38	0.02	0.00
75	0.390	3	0.33	0.05	0.01
70	0.354	2	0.36	0.00	0.02
65	0.336	3	0.31	0.07	0.03
55	0.288	3	0.27	0.03	0.07
50	0.251	2	0.25	0.04	0.13
40	0.205	2	0.13	0.04	0.25
30	0.156	2	0.00	—	—
20	0.110	2	0.00	—	—
10	0.057	2	0.00	—	—

TABLE 5 Growth rates of stems and stolons in various dilutions of seawater

Relative Seawater Strength	Day	Total Length (in mm)		% of Total Length		No. of Hydranths	Stolon Length per Hydranth (in mm)	Total Length (in mm) of New Growth		% of New Growth	
		As Stolon	As Stem	As Stolon	As Stem			As Stolon	As Stem	As Stolon	As Stem
100	0	100	28	78	22	13	7.7				
	3	448	186	71	29	54	8.3	348	158	68.8	31.2
70	0	221	47	82	18	21	10.5				
	3	674	269	77	24	98	8.9	653	222	74.8	25.2
50	0	237	29	89	11	18	11.5				
	3	333	131	72	28	49	6.8	96	102	48.5	51.5

NOTE: The lengths of stems and stolons of colonies grown in 100-percent, 70-percent, and 50-percent artificial seawater were measured at the start of the experiment and 3 days later (see text).

both the stolons and the hydranths dissociated within 24 hours. In 2×10^{-3} M potassium, many hydranths underwent regression and most of the others were unable to capture food.

In the absence of calcium ions, all of the hydranths were resorbed within 24 hours. At high calcium concentrations, a rapid cellular dissociation occurred. In 2×10^{-3} M calcium, the hydranths were unable to ingest captured *Artemia*, and growth was reduced.

At least 3×10^{-2} M magnesium is required for growth in *Bougainvillia*. In addition, contrary to the results obtained with other

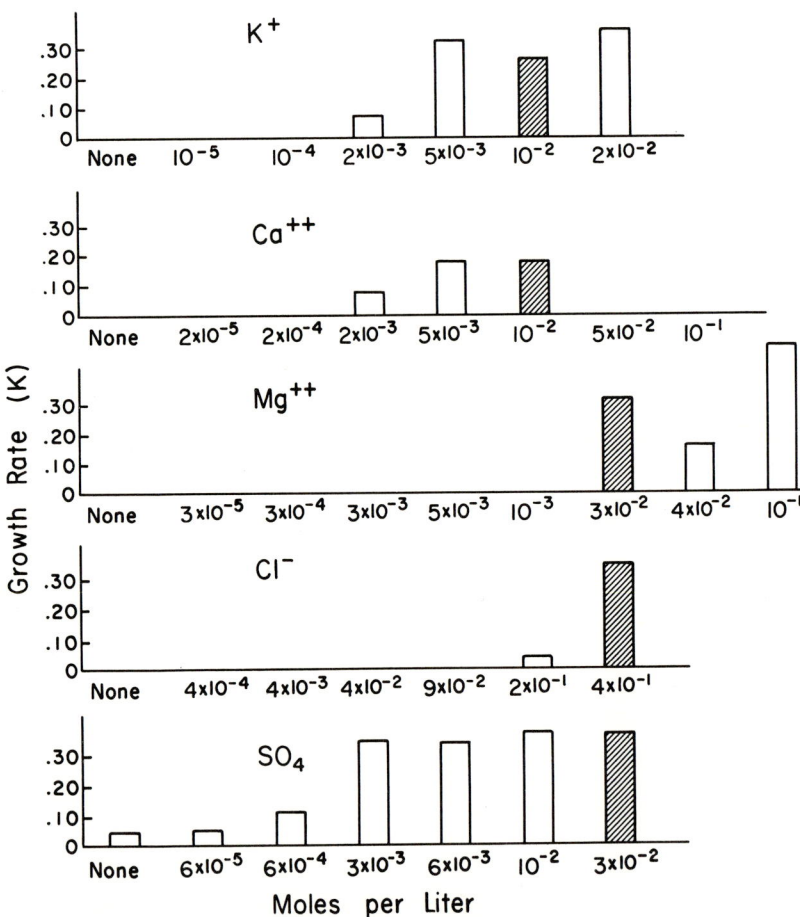

FIG. 2. Histograms of ionic requirements for growth in *Bougainvillia* sp. Hatched bars show the concentration of the ions present in the artificial seawater solution which was used as a standard culture solution.

ions, relatively high concentrations of magnesium do not inhibit growth. In one experiment a growth rate of 0.495 was obtained in 0.1 M $MgCl_2$.

In a medium containing 2×10^{-1} M chloride, colonies of *Bougainvillia* survived, but the hydranths were very small and were unable to ingest *Artemia* nauplii. Except for the colonies grown under standard conditions, no growth occurred at any other chloride concentration tested (Figure 2).

In 3×10^{-3} M sulfate, growth rates are comparable to those of the control cultures. In the absence of sulfate, however, growth continues, but only at a very low rate ($k < 0.1$). Older hydranths on these colonies regressed, while younger hydranths continued to grow and to capture and ingest food. In addition, the coenosarc of the stolons became very twisted and uneven in appearance, and often appeared to pull away from the perisarc. These abnormalities were associated only with the absence of sulfate from the medium, and were never seen in response to variations in any other environmental factor.

Temperature

The temperature tolerance of *Bougainvillia* sp. is shown in Table 6. At 11° C, all colonies died within 2 days. At 16° C, colonies remained viable with no growth occurring for 3 to 4 days. The polyps then seemed to adapt to this temperature, and a slow growth ensued. At 22° C, 30° C, and 34° C, the rate of growth was relatively fast. In a separate experiment, not shown in Table 6, seven replicate cultures at 28° C yielded a mean growth rate of 0.44, with a range of 0.14, which suggests that this temperature may be optimum for this species.

In addition, a change in colony form was observed at high temperatures. At 34° C, stolonal growth appeared to be increased and stem growth reduced; this resulted in colonies that were lower and more spread out than those grown at lower temperatures.

Nutritional Levels

The growth rate obtained when colonies were fed at 12-hour intervals was not much greater than that obtained when colonies were fed

TABLE 6 Growth rates of *Bougainvillia* sp. by temperature

Temperature (° C)	Number of Hydranths on Day:							Growth Rate (k)
	0	1	2	3	4	5	6	
11	14	7	0	—	—	—	—	0.00
16	31	31	28	33	40	52	59	0.14
22	18	24	30	51	65	90	112	0.32
30	19	40	70	91	141	180	209	0.39
34	6	12	22	31	42	76	120	0.41

at daily intervals (Table 7). On the other hand, the growth of colonies that were fed at 3-day intervals was reduced; the increased range in growth rates observed in this group suggests also that the animals were under considerable stress. Colonies that were not fed for 7 days of the experiment decreased steadily during this period in both hydranth number and in the size of individual hydranths.

DISCUSSION

The vigorous growth of *Bougainvillia* sp. under laboratory conditions indicates that the culture methods, as developed by previous investigators for freshwater polyps (Loomis 1954) and brackish-water hydroids (Fulton 1960), can also be adapted to this marine colonial hydroid. The use of a defined medium, at constant temperature, with regular feeding, offers a simple, reproducible method by which this organism can be maintained in the laboratory.

The mean growth rate of *Bougainvillia* is 0.32, a rate which is higher than the growth rate of 0.23 obtained for *Cordylophora lacustris* (Fulton 1962) and lower than the rate of 0.37 for *Hydra littoralis* (Loomis 1954) or that of 0.42 in *Chlorohydra viridissima* (Muscatine and Lenhoff 1965). However, growth rates obtained by counting numbers of hydranths may not be entirely comparable between solitary and colonial hydroids because this method does not take into account stolonal growth. Another possible limitation to these data is that, in both *Bougainvillia* and *Cordylophora*, the measurements were made on young colonies, and hence would probably not extend indefinitely. A study of mature colonies of *Bougainvillia carolinensis* indicated that the growth rates decrease with increasing age (Brock and Strehler 1963).

Growth rates for cultures of *Bougainvillia* remained relatively constant even at salinities as low as 50-55 percent that of normal seawater. However, when the salinity of the medium was reduced to 40 percent that of seawater, the mean growth rate declined to 0.13. Table 8

TABLE 7   Growth rates of *Bougainvillia* sp. by length of feeding interval

Interval of Time between Feeding (days)	No. of Hydranths/Day							Growth Rate[a] (k)
	0	1	2	3	4	5	6	
0.5	27	45	58	92	120	144	153	0.30
1.0	26	36	50	68	80	98	115	0.25
3.0	27	35	28	42	53	61	75	0.17
7.0	24	36	43	30	36	25	16	0.1

[a] Average of two experiments in which cultures were fed for 1 hour at the designated intervals.

shows the estimated concentration of each of the ionic species concerned in the artificial medium at 40 percent concentration, and the corresponding growth rates as determined by varying the ions of the medium individually (Figure 2). There are sufficient amounts of calcium, potassium, and sulfate ions at this concentration, but the concentrations of sodium chloride and of magnesium chloride were such that any or all of these ions could have become limiting. In addition, the obvious osmotic imbalance that would exist at such a low concentration of sodium chloride undoubtedly produced conditions unfavorable for continued growth at normal rates.

That the low concentration of magnesium ions was at least partially responsible for the reduction in growth seems obvious, however, because at least $3 \times 10^{-2} M$ magnesium was required for growth, and the estimated concentration of magnesium in the 40 percent artificial seawater solution was slightly below this level. Further support for the importance of magnesium to marine polyps can be adduced from the results of Kiel (1932) who found that regeneration in *Pennaria* was affected more by a decrease in the relative concentration of magnesium than by any other ionic species. Similarly, Morgulis (1909) showed that regeneration in the worm *Podarke* can be accelerated by adding small amounts of magnesium to the medium. Furthermore, although *Cordylophora* (Fulton 1962) and *Chlorohydra* (Muscatine and Lenhoff 1965) will grow slowly in the complete absence of environmental magnesium ions, the addition of small amounts of magnesium to the medium will enhance the growth of both species.

The only ionic species found to enhance the growth of *Bougainvillia*, in the sense that growth was possible in its absence but only at a

TABLE 8 Growth rates of *Bougainvillia* sp. at estimated ion concentrations

Estimated Concentration of Each Ion in Media at 40% Concentration of Artificial Seawater		Estimated Growth Rate at the Indicated Ion Concentration (k)
Salt	Moles/Liter	
NaCl	2×10^{-1}	<0.10
MgCl$_2$	2×10^{-2}	<0.10
Na$_2$SO$_4$	1×10^{-2}	0.39
CaCl$_2$	4×10^{-3}	<0.18
KCl	4×10^{-3}	<0.32

NOTE: This table compares the estimated moles per liter of five ions in the artificial seawater medium when diluted to 40-percent concentration, and the estimated growth rate at that particular ion concentration as determined by varying the ions of the media individually. The mean growth rate for cultures in media diluted to 40-percent concentration is 0.13.

considerably reduced rate, was the sulfate ion. However, the abnormalities observed in the coenosarc in the absence of sulfate suggest that this ion may be required for the continued well-being of the colony.

Changes in the form of colonies were observed at a salinity of 50 percent that of seawater, and at a temperature of 34° C. When the salinity was reduced, the rate of growth of stems was increased relative to that of the stolons, so that these colonies tended to be slightly more upright and less spread out than those grown at higher salinities. In contrast, colonies grown at 34° C tended to be more spread out, with less upright growth, than those grown at lower temperatures.

In *Bougainvillia superciliaris*, Berrill (1949) observed that, at relatively high temperatures, free stolons were formed in large numbers from lateral stems; at somewhat lower temperatures, these outgrowths gave rise to hydranths; and, at still lower temperatures, to gonophores. In the present investigation we found that, although stolonal growth increased relative to stem growth when colonies were maintained at high temperatures, the increase in stolonal growth did not occur at the expense of hydranth production. Hydranth production was, in fact, considerably higher at temperatures of 28° C, 30° C, or 34° C than it was at any lower temperatures. Furthermore, gonophore production was never observed in *Bougainvillia* sp., even at temperatures as low as 11° C, except when the cultures were neglected. It is possible, however, that the length of time during which observations were made on cultures maintained at low temperatures was too short for the appearance of gonophores to be detected. Nevertheless, it seems clear that the developmental fate of outgrowths, at least in the species of *Bougainvillia* used in this study, is not determined entirely by the temperature at which the colonies are maintained, but that some other factor (or factors) is involved. The nature of this additional factor is completely unknown.

LITERATURE CITED

Berrill, N. J. 1949. Growth and form in gymnoblastic hydroids. I. Polymorphic development in *Bougainvillia* and *Aselomaris. J. Morphology* **84**: 1–30.

Braverman, M. 1963. Studies on hydroid differentiation. II. Colony growth and initiation of sexuality. *J. Embryology and Experimental Morphology* **11**: 239–253.

Brock, M. A., and B. L. Strehler. 1963. Studies on the comparative physiology of aging. IV. Age and mortality of some marine cnidarid in the laboratory. *J. Gerontology* **18**(1): 23–28.

Crowell, S., and C. Wyttenbach. 1957. Factors affecting terminal growth in the hydroid *Campanularia. Biological Bulletin, Woods Hole* **113**: 233–244.

Fulton, C. H. 1960. Culture of colonial hydroids under controlled conditions. *Science* **132**: 473–474.

———. 1962. Environmental factors influencing the growth of *Cordylophora. J. Experimental Zoology* **151**(1): 61–78.

Hammett, F. S. 1950. Quantitative growth in *Obelia* colonies in culture. *Growth* **14**: 263-264.
Kiel, E. M. 1932. The effect of salts upon the regeneration of *Pennaria tiarella. J. Experimental Zoology* **63**: 447-455.
Loomis, W. F. 1954. Environmental factors controlling growth in *Hydra. J. Experimental Zoology* **126**: 223-234.
Morgulis, S. 1909. Contributions to the physiology of regeneration. I. Experiments on *Podarke obscura. J. Experimental Zoology* **7**: 595-642.
Muscatine, L., and H. M. Lenhoff. 1965. Symbiosis of hydra and algae. I. Effects of some environmental cations on growth of symbiotic and aposymbiotic hydra. *Biological Bulletin, Woods Hole* **128(3)**: 415-424.
Subow, N. N. 1931. *Oceanographical tables.* 208 pp. USSR. Moscow: Hydro-meteorol Committee. USSR, Oceanographic Institute USSR.

S. ARTHUR REED

CHAPTER 6

Michigan State University, East Lansing

Techniques for Raising the Planula Larvae and Newly Settled Polyps of *Pocillopora damicornis*

Planula larvae of corals, reared in controlled laboratory conditions, can prove to be valuable organisms for studying the initial stages of skeletal deposition as well as larval development. The larvae can be collected in large numbers; they readily affix themselves permanently to glass microscope slides where they can be manipulated and observed in detail. Techniques for collection, maintenance, and feeding of the larvae and polyps are described. Also included are some general comments that may be useful to those who are interested in using these organisms in research.

COLLECTION AND MAINTENANCE

Heads of the colonial coral *Pocillopora damicornis* ranging in size from about 6 inches to 18 inches in diameter were removed from the reef and transported to the laboratory in buckets of seawater. Some heads were transferred intact to aquaria containing filtered, aerated seawater; others were first broken into smaller pieces with a hammer and chisel before being transferred to aquaria. If overly crowded, the coral died and deteriorated within 18 hours.

Planula larvae (Figure 1), when present in the coral colony, would begin to be expelled by the polyps almost immediately after collection. Water in the transporting pail often contained a small number of larvae when the coral was brought into the laboratory.

Glass open-top boxes were constructed of seven microscope slides held together with cellophane tape; three slides formed the bottom and four slides formed the sides of the box. A wooden block was used to support the slides while they were being taped together.

The glass box was placed in a large finger bowl containing filtered, aerated seawater. The water level was adjusted so that the boxes were just full. The planula larvae were removed from the aquaria as they emerged from the coral head and 20–50 were placed in each of the water-filled glass boxes. Transfer was easily accomplished with little or no damage to the larvae through use of a glass pipette (Dispopets) drawn to a fine point and attached to a long rubber tube and mouthpiece.

Larvae were transferred daily into clean glass boxes with fresh seawater. If any larvae became attached to the surface of the glass slides during the 24-hour period (Figure 2), the box was carefully dismantled, the individual slides with newly affixed polyps were transferred to a container of fresh seawater, and careful observations were made thereafter on these settled larvae. A large number of slides with attached polyps were efficiently handled by using glass staining trays. The corals were illuminated for 12 hours each day by a bank of four cool, white fluorescent lamps suspended about 2 feet above the culture dishes.

FEEDING

Planula larvae, before settling, would not feed on nauplii of the brine shrimp *Artemia salina,* although they easily immobilized the nauplii with their nematocysts. After the attached polyps had developed tentacles (Figure 3), they fed readily on newly hatched *Artemia* nauplii. Extracts of

FIG. 1. Planula larva (1.5 mm) of *Pocillopora damicornis.* The oral opening is at the narrower end. The dark longitudinal stripes are due to clusters of zooxanthellae in the endoderm tissue.

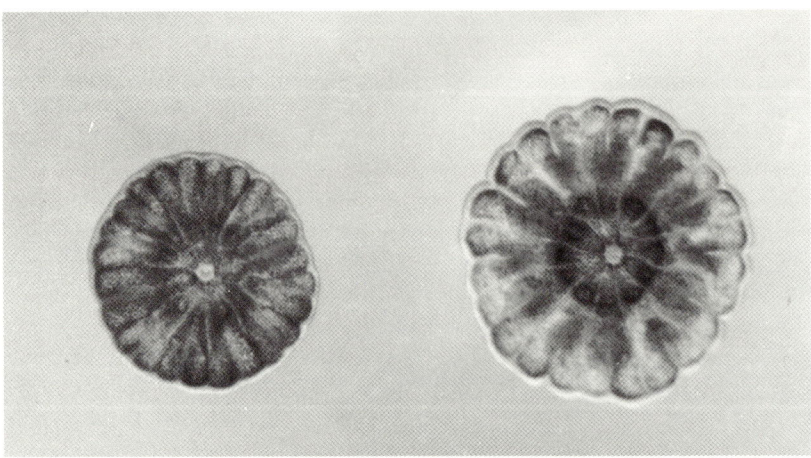

FIG. 2. Polyps (1.5 mm and 2 mm diameter) about 18 and 24 hours after settling of the larvae. Primary and secondary skeletal septa and tentacular buds are visible in the older (larger) specimen.

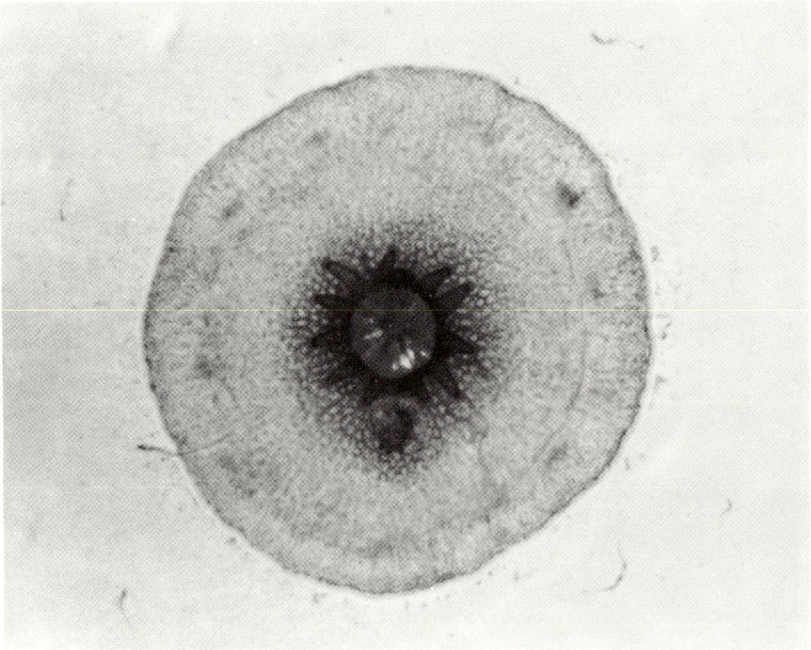

FIG. 3. Polyp (4.5 mm diameter) 11 days after settling. The first secondary polyp bud, less than 1 day old, has begun to develop.

commercially available frozen brine shrimp adults elicited a strong feeding response in the settled coral polyps. The coral also readily accepted such animal tissue as small chunks of fresh or frozen fish muscle or liver and pieces of liver from mice and geckos. The chunks of muscle and liver were placed directly on the tentacular surface of the polyp with fine tweezers; hence, experiments in which the polyps are fed pieces of radioactive liver should be feasible. A polyp which has been fed *Artemia* nauplii off and on for 61 days is shown in Figure 4.

OBSERVATIONS AND COMMENTS

A number of workers have described in detail various aspects of maintenance, behavior, anatomy, and physiology of a variety of coral larvae (see "References on Coral Larvae"). A few observations not previously recorded are described here.

Planula larvae could often be located within individual polyps of the parent colony with the aid of a dissecting microscope. These larvae could

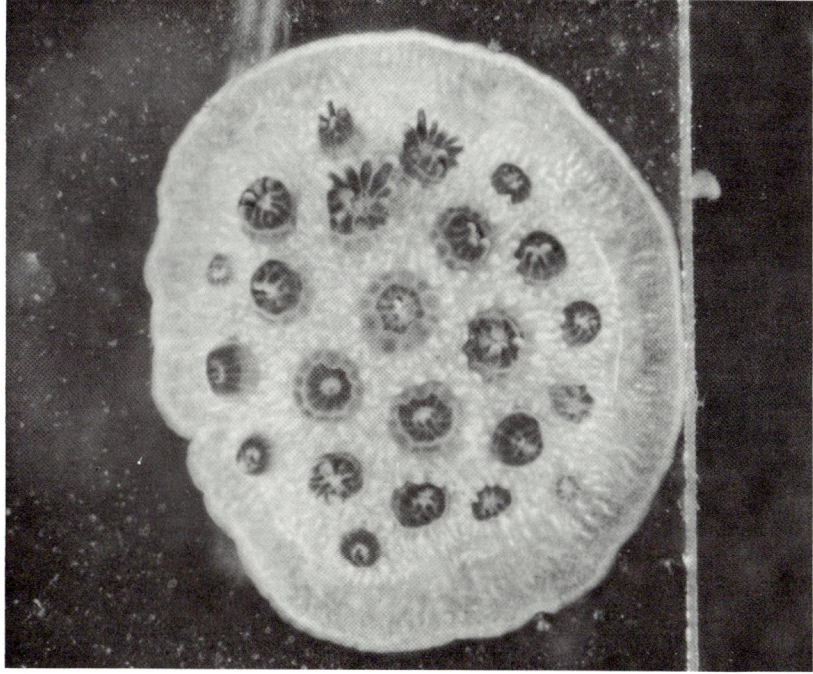

FIG. 4. A colony of polyps (11 mm diameter) 61 days after settling of a single planula larva. The central polyp, contracted in this figure, was produced first. A total of 31 polyps, not all visible, are in various stages of development.

be forced from the polyp by gently scraping the tissue with a teasing needle or glass probe. A few larvae were removed using a pipette alternately to blow and suck jets of water over the oral surfaces of the polyps. These actions gradually drew the larvae from the gastrovascular cavities with little apparent damage either to the parent polyps or the larvae.

The glass slides, which became coated with algae and bacteria, were cleaned about every third day by using a small camel-hair brush under a dissecting microscope. Such care, though often tedious to the researcher, caused little damage to the affixed polyps.

Occasionally an unusual Siamese-twin-like form of the larva was collected (Figure 5). Five such larvae observed had little difficulty in moving around, and settled on glass slides, always attaching at the fused basal end. In every case two separate polyps developed, each having its own set of tentacles, but a common basal disc (Figure 6).

Attempts were made to collect larvae during November 1967, and April, June, July, and August of 1968. Coral heads were brought from the reef at intervals of approximately 3 days during these months. In November and April planulae were expelled only during a 9-day interval just before and after the new moon. During the three summer months larvae were obtained at every collection, usually in numbers upwards to 150 per coral head. Marshall and Stephenson (1933) in studies done on the Great Barrier Reef, were able to correlate expulsion of planulae in

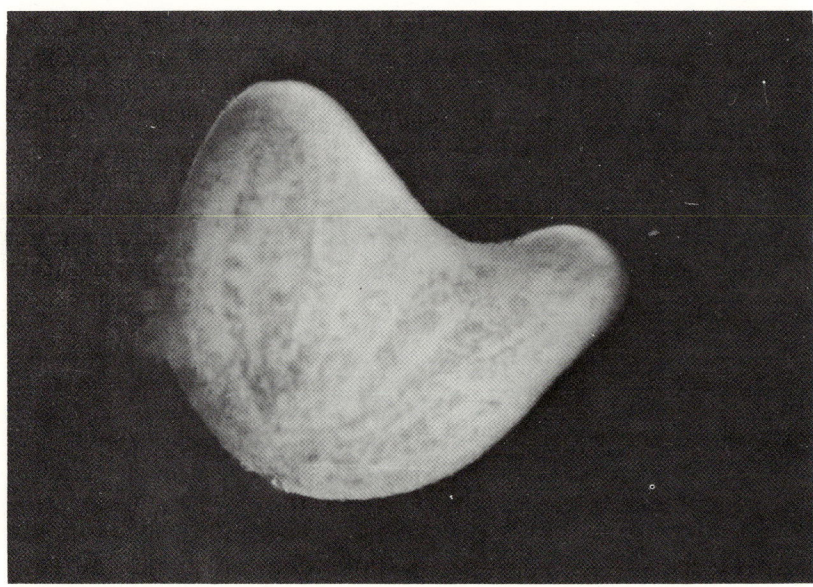

FIG. 5. An unusual form of the planula larva, fused at the blunt aboral end.

Techniques for Raising Larvae and Polyps of Pocillopora damicornis

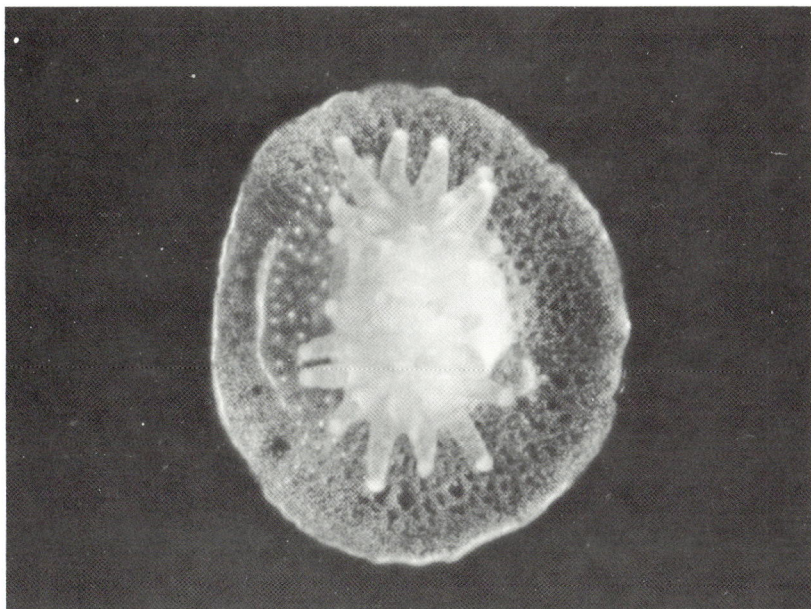

FIG. 6. Polyp (3 mm diameter) 9 days after settling of unusual planula larva (shown in Fig. 5). Two complete sets of tentacles have developed around the two oral openings.

Pocillopora bulbosa with the new moon from December to April and with the full moon from July to November. My limited data roughly correspond to these expulsion periods. The most favorable months for obtaining larvae in large numbers in Hawaii appear to be the summer months.

REFERENCES ON CORAL LARVAE

Abe, N. 1937. Post-larval development of the coral *Fungia actiniformis* var. *palawensis* Doderlein. *Palao Tropical Biological Station Studies* 1: 73–93.

Atoda, K. 1947. The larval and postlarval development of some reef-building corals. I. *Pocillopora damicornis cespitosa* (Dana). *Scientific Reports, Tohoku University*, series 4 (Biology), 18: 24–47.

———. 1951. The larval and postlarval development of reef-building corals. III. *Acropora bruggemanni* (Brook). *J. Morphology* 89: 1–16.

———. 1951. The larval and postlarval development of reef-building corals. IV. *Galaxea aspera* (Quelch). *J. Morphology* 89: 17–36.

Boshma, H. 1929. On the postlarval development of the coral *Meandra areolata* (L.). *Carnegie Institution of Washington Publications*, 391: 131–147.

Edmondson, C. H. 1929. *Growth of Hawaiian corals.* Bernice P. Bishop Museum, bulletin 58. Honolulu.

———. 1946. Behavior of coral planulae under altered saline and thermal conditions. Bernice P. Bishop Museum, Occasional Papers 18, pp. 283–304.

Kawaguti, S. 1941. On the physiology of coral reefs. V. Tropisms of coral planulae, considered as a factor of distribution of the reefs. *Palao Tropical Biological Station Studies* **2**: 319–328.

Marshall, S. M. 1932. Notes on oxygen production in coral planulae. *Scientific Reports of the Great Barrier Reef Expedition* **1**: 253–258.

Marshall, S. M., and T. A. Stephenson. 1933. The breeding of reef animals. Part I. The corals. *Scientific Reports of the Great Barrier Reef Expedition* **3**: 219–245.

Needham, J. G., ed. 1959. *Culture methods for invertebrate animals.* New York: Dover Publications.

Stephenson, T. A. 1931. Development and the formation of colonies in Pocillopora and Porites. Part I. *Scientific Reports of the Great Barrier Reef Expedition* **3**: 113–134.

Stephenson, T. A., and A. Stephenson. 1933. Growth and asexual reproduction in corals. *Scientific Reports of the Great Barrier Reef Expedition* **3**: 167–217.

Vaughan, T. W. 1910. The recent Madreporaria of southern Florida. *Carnegie Institution of Washington Yearbook* **9**: 135–144.

———. 1911. The Madreporaria and marine bottom deposits of southern Florida. *Carnegie Institution of Washington Yearbook* **9**: 135–144.

PART 2

FEEDING BEHAVIOR, FOOD TRANSPORT, AND METABOLISM

HOWARD M. LENHOFF
University of California at Irvine

CHAPTER 7

Research on Feeding, Digestion, and Metabolism in Coelenterates: Some Reflections

Twenty years ago investigations on feeding in coelenterates were carried out almost exclusively by students of natural history and behavior. Experiments on coelenterate digestion received attention mostly from those employing cytological techniques and from a few marine biologists who assayed for lytic enzymes and the pH optima of those enzymes. Research on the metabolism and biochemistry of coelenterates was practically nonexistent. (For a comprehensive review of recent work on these subjects, see Lenhoff 1968*a*.)

Yet, although much work on coelenterates, like that on most invertebrates except insects, has not been part of the march of modern biology, some hopeful signs have emerged. Zoologists here and there have begun using modern tools to investigate coelenterates, and a few biochemists and physiologists have begun using coelenterates as experimental organisms. As a result, chemical studies on feeding behavior, digestion, and metabolism have gotten off the ground–but not much farther. Too few animals have been investigated, and too few modern techniques have been applied.

It is of interest that much of the current work on coelenterates was done using hydras. Even in this volume, for example, we describe a number of investigations with these freshwater animals which were carried out at a marine laboratory in the middle of the Pacific Ocean. The reason for the heavy emphasis given to hydras, despite the large variety and number of coelenterates available, has been expounded on in Chapter 1.

As a consequence of the successful laboratory experimentation upon hydras, these animals have become useful prototypes for those interested in the biology of marine coelenterates. Discoveries made with hydras have unlocked major areas in coelenterate biology; with the background information thus gained, the route to fruitful experimentation with marine organisms, most of which are less hardy in the laboratory, became shorter and more direct. For example, the recent experimental work on the feeding behavior, digestion, symbiosis, nematocyst chemistry, and enzyme activities of hydra all preceded and provided the basis for much of the work of a biochemical nature described in this volume.

Students of coelenterate biology may wish to learn about the circuitous and indirect routes by which many of these experiments actually originated. Such personal reflections as are presented in the remainder of this introductory chapter serve to illustrate a key point made in Chapter 2: that is, through a broad organismic approach, previously unsuspected phenomena may be revealed.

CHEMICAL CONTROL OF FEEDING BEHAVIOR

The activation of the feeding response of hydras by reduced glutathione is now generally known by biologists, and has been reviewed many times (Lenhoff 1961*b*, 1965, 1968*a, b*). It is perhaps one of the clearest examples of a feeding response being controlled by a single species of a small molecule. And more recently, research on the glutathione-hydra system has evolved into a quantitative study of a receptor site (Lenhoff 1969) and of interactions of different behavioral receptor-effector systems (Rushforth 1965; Blanquet and Lenhoff 1968; Lenhoff 1968*b*). Yet had we been sensory physiologists or behavioral biologists in 1954, we most certainly would not have selected a hydra as the experimental animal for such studies. Little was known of the physiology and nervous system of these animals, and it did not seem amenable to quantitative research. Loomis's initial reason for studying feeding in *Hydra littoralis*—to determine the nutritional requirements of the animal—was to gain further control over his experimental system for developmental studies. Loomis soon learned, as others did before him, that *Hydra* ingests particulate food, and that food extracts stimulate a feeding response. Before he could investigate the composition of *Hydra*'s food, and before he could have precise control over the *Hydra*'s ingestion of particles, Loomis had to identify the compound in food extracts that induced a feeding response. Once he proved that reduced glutathione was the specific activator of feeding, new possibilities for chemoreceptor and behavioral research opened (Loomis 1955).

Loomis's finding was broadened when Fulton (1963) discovered that the colonial hydroid *Cordylophora lacustris* gave a feeding response specifically to proline and to some of its analogs, but not to glutathione. Hence, it appeared that, as Loomis (1955) had predicted in his original article: "The chemical mediator involved [in the feeding reactions of other coelenterates] may consist of glutathione in certain cases, as in hydra, or may consist of some other cell constituent that functions in a similar manner."

Recent research, including that carried out by the students in this program, has led to the discovery of many feeding activators, most of them amino acids, which affect specific coelenterates. Some of these compounds and the responding coelenterates are listed in Table 1. It

TABLE 1 Natural substances that activate phases of coelenterate feeding behavior

Substance	Organism	Reference
Proline	*Cordylophora lacustris*	Fulton 1963
	Pennaria tiarella	Pardy, this volume
	Cyphastrea ocellina[a]	Mariscal, this volume
Glutathione	*Hydra*	Loomis 1955
	Physalia physalis	Lenhoff and Schneiderman 1959
	Nanomia cara	Mackie and Boag 1963
	Campanularia flexuosa	Lenhoff and Schneiderman 1959
Valine	*Boloceroides* sp.	Lindstedt, this volume
Glutamine	Acontiate sea anemone[b]	Smith and Lenhoff, unpublished
Alanine	*Aurelia* sp.	Kauffman and Muscatine, unpublished
Tyrosine[c]	*Hydra*	Blanquet and Lenhoff 1968

[a]*C. ocellina* also responds to higher concentrations of reduced glutathione.
[b]Found on a sargassum weed in Biscayne Bay, Florida.
[c]Tyrosine activates an enteroreceptor in the gut of hydra. When, at the same time, the external glutathione receptors are activated, the hydra give a "neck" response (Blanquet and Lenhoff 1968).

seems reasonable to expect that, as more coelenterates are investigated, a whole spectrum of amino acids, peptides, and possibly other substances may be found to function as specific activators of feeding.

CHEMISTRY OF INTRACELLULAR DIGESTION

Because coelenterates combine both extracellular and intracellular digestion, this function has long interested biologists. Yet research on coelenterate digestion has not progressed much beyond Metschnikoff's classical cytological studies of the late nineteenth century. Most biochemical research on coelenterate digestion, which has centered on surveys for activities of lytic enzymes and their pH optima, has received a fair amount of criticism (see Lenhoff 1968a).

Perhaps one of the major roadblocks to biochemical investigations of coelenterate digestion is the technical difficulty involved in analyzing the catabolic changes occurring within the food vacuole wherein most of the degradation of food into assimilable molecules takes place. Unlike the gut, the food vacuole could not be sampled to analyze for metabolic intermediates. On top of that, simple chemical analyses of gastrodermal tissues do not distinguish between materials within the food vacuoles and similar materials in the cells containing those vacuoles.

The work presented by Murdock in this volume shows how this roadblock was overcome in the investigation of protein digestion in a sea anemone and in a hydra. The basic technique was relatively simple and straightforward. The animals were induced to ingest tissue labeled previously with radioactive amino acids (most of the label was in the tissue protein). Once the labeled food was ingested, kinetic analyses were made of the distribution of amino acids, large peptides, and proteins. These analyses were made of the material in the lumen of the gastrovascular cavity and of the coelenterate tissues which contained the food vacuoles. The results of such studies are described in detail (Lenhoff 1961a, 1968a; Murdock, this vol.). In summary, through such studies it was possible to analyze the rate and efficiency of gastrodermal phagocytosis; the extent, rate, and place of protein digestion; the existence of alcohol-soluble peptides (or small proteins) as intermediates and possible storage materials within the food vacuole; and the rate of transfer of products of digestion to cells not containing food vacuoles.

The relative simplicity of these techniques and their application to studies of intracellular digestion were not apparent at the outset. As might be expected from the theme of this discussion, my initial investigations (Lenhoff 1961b) along these lines originated indirectly from another research problem. At the Biophysics Division of the Carnegie Institution of Washington, I was searching for a means by which hydra nematocysts could be labeled with large amounts of ^{14}C-proline. To my disappointment only very small amounts of the label were taken up, and we could not be sure whether that was taken up by the hydra directly or by microorganisms adhering to the animal's surface.

Because large amounts of radioactive amino acids of high specific activity were somewhat difficult to come by at that time, I took advantage of some heavily labeled "particulate" material left over from a colleague's experiment—the remains of a carcass of a ^{35}S-labeled mouse. The hydra ingested and retained bits of radioactive mouse tissue and became heavily labeled (Lenhoff 1958). Before continuing with the labeling studies per se, it was necessary to determine the efficiency of uptake of labeled materials, the degree of retention, and the fate of the label. In this manner, our investigations of the mechanisms of intracellular digestion originated. To date these methods have been used only on studies of protein digestion with virtually no work done as yet on the mechanism of digestion of other major food substances.

These techniques, however, have been modified and applied to a number of other studies: (a) to determine the degree to which endosymbiotic algae affect the metabolism of their host cells (Muscatine and Lenhoff 1965); (b) to determine the degree to which CO_2 produced in coral metabolism is used for mineralization (Pearse, this vol.); (c) to determine possible dependency of endosymbiotic algae on organic material released from host cells (Cook, this vol.); (d) to investigate the distribution of ingested food through the various parts of a colonial hydroid (Rees, this vol.); and (e) to determine whether or not bacteria ingested by corals are assimilated (DiSalvo, this vol.).

There are obviously other uses for these methods in studies of coelenterate metabolism, physiology, and development. Furthermore, they should be readily extended to investigations of other organisms that feed on large particles or tissues (e.g., flatworms, nudibranches). Improvements of the methods are now being devised independently by C. Cook and B. Barzansky, who are preparing heavily labeled *Artemia* nauplii that can be fed easily and efficiently to coelenterates and to other aquatic organisms.

METABOLISM AND BIOCHEMISTRY

Although coelenterates are regarded now as being useful for studies of development, behavior, and symbiosis, they do not appear to biochemists as being objects worthy of "serious" biochemical investigation.

This attitude seems strange, indicative of short memories, because one of the most exciting chapters of present-day biochemistry had its origins with research on coelenterates; i.e., one of the first pigments most likely to be classified today as a cytochrome was found in a sea anemone.

I refer to the classical investigations of C. A. MacMunn, who first discovered what are now known as cytochromes in a host of tissues and organisms (1886). He called them myohaematins and histohaematins. As told in many a biochemistry textbook, his discoveries were discredited by the unflagging argumentation of the well-known and influential Hoppe-Seyler. It was not until Keilin in 1926 rediscovered cytochromes that the significance of MacMunn's work became recognized.

Although MacMunn's 1886 survey is widely referred to today, little mention is given to actinohaematin, a substance which he describes in 1885 as "a respiratory coloring matter . . . present in *Actinia mesembryanthemum, Bunodes crassicornis,* and other Actinae. That it must be respiratory is shown by the fact that one of its decomposition products is capable of existing in a state of oxidation and reduction." Only after he showed pigments with such properties in many forms (1886) did his argument gain strength. But much of his initial impetus came from experiments on coelenterates.

My own experiences with the metabolism and biochemistry of coelenterates, although not extensive, have been full of surprises. I will give two examples related to the work described in this volume.

Collagens. Our investigations on the chemistry of nematocysts originated from our curiosity about a "white sediment" always observed at the bottom of homogenates of dense suspensions of hydra. Further purification and analyses of the white sediment, which consisted mostly of capsules of broken discharged nematocysts, showed it to be extremely rich in hydroxyproline and proline (Lenhoff, Kline, and Hurley 1957). We thought the capsule protein to be an unusual kind of collagen though, at that time, collagens were considered primarily to be extracellular fibrous material found almost exclusively in the vertebrates. In the late 50s a host of collagenlike proteins was found in many invertebrates (see Piez and Gross 1959). Our interest in this capsular protein waned because we could not solubilize it as could be done with vertebrate collagens.

We began to reinvestigate the capsule protein 10 years later, this time using preparations from the sea anemone *Aiptasia pallida.* Dr. Richard Blanquet, extending an observation of Yanagita (1959), showed that nematocyst capsules dissolve in disulfide reducing agents (Blanquet and Lenhoff 1966). These findings posed a paradox because all collagens were thought to be devoid of sulfur amino acids, and, furthermore, insoluble proteins that dissolved in disulfide reducing agents were thought to be keratins. This problem was resolved when we showed that the nematocyst capsule, when dissolved, gives only one protein band on disc electrophoresis (Blanquet and Lenhoff 1966). Thus, it appears that the coelenterate nematocyst capsule contains as a major component a single type of collagen protein linked by disulfide bonds. In this volume Mariscal shows that these findings on *Aiptasia* nematocysts can be extended to the nematocysts from a number of Hawaiian coelenterates.

Disulfide bonds in collagen from another invertebrate were described about the same time by McBride and Harrington (1965), who were investigating the cuticle from *Ascaris,* a nematode. It seems clear that, as further work with collagens from invertebrates occurs, the sulfur amino acids, previously thought to be contaminants of "impure" collagens, may also be shown to be involved in forming disulfide bridges. Hence, we have here examples of how investigations of invertebrate material may alter concepts derived solely from work on vertebrate tissues.

It was not a distant jump from investigations of nematocyst collagens to mesogleal collagens. Gosline (this vol.) presents an excellent example of how the kinetics of *in vivo* hydroxylation of protocollagen to collagen can be demonstrated, using a sea anemone.

Enzyme pathways. It might generally be assumed that the major biochemical pathways of most metazoans do not differ drastically from one another. At least we thought so, and felt that a survey for enzyme activities in coelenterates would provide us with no more than a catalog of enzymes. Hence, although my original training was in enzymology, my laboratory refrained from embarking on such projects.

Only when Dr. C. Rutherford was investigating possible roles of metabolic gradients in *Hydra littoralis* did we note deviations in some coelenterates from so-called normal metabolic pathways. In a careful study, Rutherford initially noted differences in distribution of the activity of glucose-6-phosphate dehydrogenase (G6PDH—the first enzyme in the pentose phosphate shunt) along the body tube of hydra. Rather than conclude that the distribution of G6PDH which he had observed reflected the distribution of the pentose phosphate pathway, Rutherford assayed for the second enzyme in the shunt, 6-phosphogluconate dehydrogenase (6PGDH). Because these two enzymes normally occur in about 1:1 ratio in tissues having the shunt, we were astonished when the results conclusively demonstrated that there is absolutely no detectable 6PGDH activity in hydra, although Rutherford had previously shown that hydras are particularly rich in G6PDH. He assayed six more species of hydras and found the same result.

These findings remained perplexing until Powers (this vol.) extended Rutherford's experiments to 21 species of coelenterates. Powers found that all of the six species of hydroids assayed also lacked 6PGDH activity. In addition, two of the three species of scyphozoans assayed also lacked the enzyme. On the other hand, extracts of all anthozoans assayed had significant 6PGDH activity.

I wish to emphasize here that there is no reason to assume that the metabolic pathways of coelenterates (and of other lower forms) have to be identical to those of higher organisms. Only when we investigate a greater number of organisms will we be able to understand fully the range of metabolic activities of which the living cell is capable.

LITERATURE CITED

Blanquet, R. S., and H. M. Lenhoff. 1966. A disulfide-linked collagenous protein of nematocyst capsules. *Science* **154**: 152–153.

———. 1968. Tyrosine enteroreceptor of hydra: Its function in eliciting a behavior modification. *Science* **159**: 633–634.

Fulton, C. 1963. Proline control of the feeding reaction of *Cordylophora. J. General Physiology* **46**: 823–837.

Lenhoff, H. M. 1958. Feeding and digestion. *Yearbook of the Carnegie Institution of Washington* **57**: 157–160.

———. 1961a. Digestion of protein in *Hydra* as studied using radioautography and fractionation by differential solubilities. *Experimental Cell Research* **23**: 335–353.

———. 1961b. Activation of the feeding reflex in *Hydra littoralis.* In *The biology of hydra and of some other coelenterates: 1961,* H. M. Lenhoff and W. F. Loomis, eds., pp. 203–232. Coral Gables: University of Miami Press.

———. 1965. Some physicochemical aspects of the macro- and micro-environments surrounding hydra during activation of their feeding behavior. *American Zoologist* **5**: 515–524.

———. 1968a. Chemical perspectives on the feeding response, digestion, and nutrition of selected coelenterates. In *Chemical zoology,* vol. 2, M. Florkin and B. Scheer, eds., pp. 157–221. New York: Academic Press.

———. 1968b. Behavior, hormones, and hydra. *Science* **166**: 434–442.

———. 1969. pH profile of a peptide receptor. *Comparative Biochemistry and Physiology* **28**: 571–586.

Lenhoff, H. M., E. Kline, and R. Hurley. 1957. A hydroxyproline-rich, intracellular, collagen-like protein of *Hydra* nematocysts. *Biochimica et biophysica acta* **26**: 204–205.

Lenhoff, H. M., and H. A. Schneiderman. 1959. The chemical control of feeding in the Portuguese man-of-war, *Physalia physalis* L. and its bearing on the evolution of the Cnidaria. *Biological Bulletin, Woods Hole* **116**: 452–460.

Loomis, W. F. 1955. Glutathione control of specific feeding reactions of hydra. *Annals, New York Academy of Sciences* **62**: 209–228.

Mackie, G. O., and D. A. Boag. 1963. Fishing, feeding and digestion in Siphonophores. *Pubblicazioni della Stazione zoologica di Napoli* **33**: 178–196.

McBride, D. W., and W. F. Harrington. 1965. Evidence for disulfide cross-linkages in an invertebrate collagen. *J. Biological Chemistry* **240**: PC 4545–4547.

MacMunn, C.A. 1885. XI Observation on the chromatology of *Actiniae, Trans. Royal Society* (London) **176**: 641–663.

———. 1886. VI Researches on myohaematin and the histohaematins. *Trans. Royal Society* (London) **177**: 267–298.

Muscatine, L., and H. M. Lenhoff. 1965. Symbiosis of hydra and algae. II. Effects of limited food and starvation on growth of symbiotic and aposymbiotic hydra. *Biological Bulletin, Woods Hole* **129**: 316–328.

Piez, K. A., and J. Gross. 1959. The amino acid composition and morphology of some invertebrate and vertebrate collagens. *Biochimica et biophysica acta* **34**: 24–39.

Rushforth, N. B. 1965. Inhibition of contraction responses in *Hydra*. *American Zoologist* **5**: 505–513.

Yanagita, T. M. 1959. Physiological mechanism of nematocyst responses in sea anemone. I. Effects of trypsin and thioglycolate upon the isolated nematocysts. *Japanese J. Zoology* **12**: 361–375.

ROSEVELT L. PARDY
University of Arizona, Tucson

CHAPTER 8

The Feeding Biology of the Gymnoblastic Hydroid *Pennaria tiarella*

The means by which cnidarians capture and ingest their prey has periodically interested workers in coelenterate biology. The name cnidaria itself relates to the nematocysts—structures, common to all members of this phylum, which assist in food capture. Despite the enormous amount of research conducted on nematocysts, there is little knowledge, except for the work of Ewer (1947), of the exact function of the different kinds of nematocysts.

In the past decade, mainly because of the work of Loomis (1955), interest in cnidarian feeding has shifted to the chemical control of feeding responses. Loomis showed that reduced glutathione was the specific chemical agent responsible for eliciting a feeding response in *Hydra littoralis*. Since that time, a number of other workers have stimulated the feeding responses of various cnidarians with reduced glutathione (Lenhoff and Schneiderman 1959) or with some other small organic molecule (Fulton 1963; Lindstedt, this vol.).

For a number of reasons it seemed advantageous to study feeding in the colonial gymnoblast *Pennaria tiarella* Agassiz, one being its ready availability in Hawaii. In addition, *Pennaria* has two different kinds of tentacles, filiform and capitate, each of which has its own peculiar array of nematocysts. This natural division allowed us to study the action of the nematocysts present in the filiform tentacles in one experiment, and the action of those in the capitate tentacles in another.

Another advantage which led me to select *Pennaria* for study was its relationship to another colonial gymnoblast, *Cordylophora lacustris*. Since *Cordylophora* gives a feeding response to proline, I wanted to determine if *Pennaria* behaved similarly.

The results of these experiments on the employment of nematocysts and the identification of a feeding activator in *Pennaria tiarella* are the subjects of this chapter.

An earlier version of this chapter appeared as "The feeding biology of the gymnoblastic hydroid, *Pennaria tiarella*" by Rosevelt L. Pardy and Howard M. Lenhoff, in the *Journal of Experimental Zoology* **168** (2): 197–202.

MATERIALS AND METHODS

Pennaria tiarella is a colonial gymnoblastic hydrozoan. The colony consists of a central stem (up to 10 cm in length) from which emerge side branches bearing individual hydranths. Lower branches have as many as eight hydranths while the most distal branches have only one. A fully developed hydranth consists of an oral cone studded with 10-18 knobbed capitate tentacles. From the base of the cone emerge 9 to 13 filiform tentacles (Figure 1).

Colonies of *P. tiarella* with attached stolons were collected from encrusted docks in front of the marine station and attached to 7" by 7" squares of metal chicken wire. These squares were kept in fresh running seawater. Preceding an experiment, a square with the attached colony was removed from the fresh seawater and placed in an aquarium containing Instant Ocean (Aquarium Systems, Inc., 1450 East 289 Street, Wickliffe, Ohio) for 24 hours. This was done to remove any organic ions present in seawater, and to assure the absence of food. Hydranths from these fasted animals were used for the feeding experiments. It is important that this procedure be followed to assure the well-being of the colony. Otherwise the hydranths may begin to regress, and thus give erratic results.

For bioassay of a chemical mediator of feeding, a stem containing three to five hydranths was removed from a colony and placed in 20 ml of a test solution contained in a 5-cm diameter petri dish. Test solutions prepared in Instant Ocean consisted of proline, pipecolic acid, methionine, phenylalanine, tryptophan, histidine, glutathione, or valine—all at 10^{-3} to 10^{-6} molar concentrations. The amino acids were purchased from Calbiochem. Also tested was a seawater extract of the brine shrimp

FIG. 1. *Pennaria tiarella.* Left, colony. Right, hydranth. *C,* capitate tentacles; *O,* oral cone; *F,* filiform tentacle.

Artemia salina. This extract was prepared by homogenizing a dense suspension of *Artemia* nauplii in Instant Ocean solution in a tissue grinder, centrifuging at 18,000 × *g* for 30 minutes, and then removing the clear supernatant layer. This extract was diluted 1:100, 1:1,000, 1:10,000 with artificial seawater. A 1:1,000 dilution contained about 0.003 mg/ml protein nitrogen as measured by the method of Lowry et al. (1951).

To record the response of the hydranths, the scoring system of Fulton (1963) was adopted: a complete reaction including cone bending and mouth opening was recorded as (++); cone bending but no mouth opening as (+-); and no response as (-). In addition, (++) was used to indicate weak cone bending and slow mouth opening. Also recorded were the times the mouth opened after the test solution was added and the average time the mouth remained open.

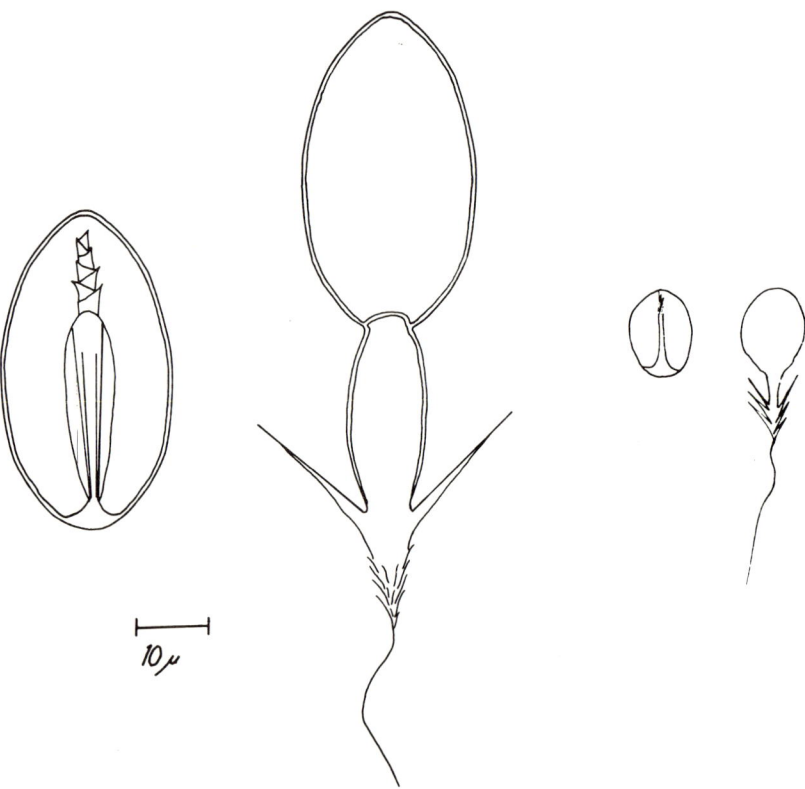

FIG. 2. *Pennaria tiarella.* Left, large stenotele from capitate tentacle; center, nematocyst; right, small stenotele from filiform tentacle.

RESULTS AND DISCUSSION

Role of Tentacles and Nematocysts in Food Capture

A typical feeding response of *Pennaria tiarella* to live food was elicited using *Artemia* nauplii. A swimming nauplius would contact one of *Pennaria*'s filiform tentacles and be made immobile by the discharged nematocysts, but at this point it did not stop all body movements. Within 2 seconds after its capture, the oral cone of the hydranth bent toward it. The nauplius at this time was contracted by one or more of the cone's capitate tentacles. Almost immediately upon contact with the capitate tentacles, the nauplius ceased to move. The mouth of the hydranth, which had been slowly opening since the nauplius was captured, pressed against the captured prey and engulfed it.

Microscopic examination of captured nauplii revealed that their appendages and antennae were entangled with numerous desmonemes, and their exoskeletons were penetrated with both large and small stenoteles. Examination of the hydranth's filiform tentacles showed them to contain numerous small desmonemes, atrichous isorhizas, and small stenoteles. In contrast, each capitate tentacle possessed five to six very large stenoteles and many atrichous isorhizas. The relative size of the filiform and capitate stenoteles is shown in Figure 2. (No doubt it is the large stenoteles from the capitate tentacles which are responsible for the painful effects of *Pennaria* on humans.)

To determine further the independent roles that the filiform and capitate tentacles play in the capture of prey, a hydranth was removed from a colony and cut so that the ring of filiform tentacles was separated from the oral cone with its staggered array of club-shaped capitate tentacles. The two components were placed in a few milliliters of seawater in separate depression slides. To each was added an *Artemia* nauplius. After contacting the filiform tentacles, the nauplii struggled as long as 7 minutes before they stopped moving. Numerous small stenoteles were found to have penetrated the nauplii's exoskeletons. On the other hand, those nauplii contacting the knob of the capitate tentacles were completely immobilized within 6 seconds. Examination of these nauplii revealed penetration of the exoskeleton with 8 to 12 large stenoteles. These observations suggest that the large stenoteles from the capitate tentacles inject more toxin, and hence play the major role in immobilizing large prey. That the nauplius gradually succumbed on contact with the filiform tentacles suggests that there is a cumulative effect of toxin injected from the numerous small stenoteles which were discharged into the struggling prey.

Chemical Mediation of the Feeding Response

When hydranths of *Pennaria tiarella* were placed in seawater containing an aqueous extract of *Artemia*, (*a*) the oral cone bent back and

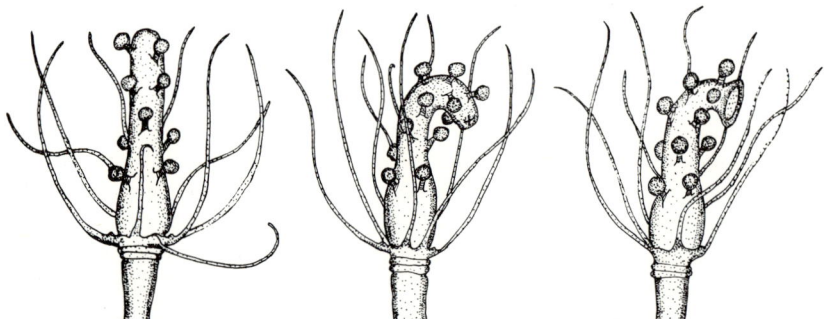

FIG. 3. Feeding response in *Pennaria tiarella*. Left, normal posture; center, cone bending; right, mouth opening.

forth; (b) the mouth opened (the cone continued to bend while the mouth was open) (Figure 3). These two distinct phases we call the feeding response. Cone bending may also be brought about by agitating the colony (Josephson 1961), but such bending was not accompanied by mouth opening. Further, when about 0.05 ml of shrimp extract was carefully added to a quiescent colony, it initiated cone bending; thus the initiation of cone bending does not require a mechanical stimulus.

The time required for the mouth to open depended upon the concentration of shrimp extract. The higher the concentration, the more quickly the mouth opened (Table 1).

When the hydranth was placed in a dilute shrimp extract, the lips of the bending cone occasionally came in contact with the bottom or wall of the container. When this occurred, the mouth spread over the surface in an apparent attempt to engulf it. A similar spreading response has been reported to occur in *Hydra littoralis* (Loomis 1955) and in *Physalia* (Lenhoff and Schneiderman 1959).

Because dilute solutions of the shrimp extract elicited a feeding response in *Pennaria*, I carried out some experiments to determine if a specific substance present in the extracts was the active component. First I tested dilute solutions of reduced glutathione and of proline, because these compounds are known to be specific feeding activators of two other

TABLE 1 The effect of shrimp extract on mouth opening behavior in *P. tiarella*

Number of Hydranths	20	25	25
Ratio of Shrimp Extract to Seawater	1:100	1:1,000	1:10,000
Time Required for Mouth to Open (min)	<1	3	>5
Average Time Mouth Remained Open (min)	6	2	1 ± 0.5

TABLE 2   Effects of proline, pipecolic acid, and glutathione on feeding behavior in *P. tiarella*

Compounds	Concentration (molarity)				
	10^{-3}	10^{-4}	10^{-5}	10^{-6}	10^{-7}
Proline	++	++	++	$\underline{++}$	—
Pipecolic Acid	++	++	—	—	—
Reduced Glutathione	—	—	—	—	—

NOTE: ++, cone bending and mouth opening; $\underline{++}$, weak cone bending and slow mouth opening; —, no response.

gymnoblasts–glutathione for *Hydra littoralis* (Loomis 1955), and proline for the colonial *Cordylophora* (Fulton 1963). Table 2 shows that proline was active at concentrations as low as 10^{-5} M to 10^{-6} M, whereas reduced glutathione did not elicit a response. Pipecolic acid, a six-membered ring analog of proline shown by Fulton (1963) to be active on *Cordylophora*, also elicited a positive response in *Pennaria*. Pipecolic acid evoked approximately one-tenth as much activity as proline in both *Cordylophora* and *Pennaria*.

In addition to glutathione, the following compounds were inactive: methionine, reported to elicit extrusion of mesenterial filaments in corals (Goreau 1961); valine, the feeding activator of the sea anemone *Boloceroides* (Lindstedt, this vol.); and histidine, tryptophane, and phenylalanine, three cyclic amino acids. In every case where no response occurred, the hydranths were subsequently found to respond to a dilute extract of *Artemia* nauplii. Hence, the inactive compounds were not toxic to the *Pennaria* hydranths.

The response to proline by *Pennaria* appeared in every way to be the same as the response elicited by the shrimp extract. In testing the effects of several proline concentrations, it was observed that the lower the concentration, the longer the time required for the mouths to open (Table 3); whereas with the higher concentrations of proline, the length of time the mouths remained open increased. I estimate from these data that the concentration of proline eliciting a half maximal mouth-opening response is approximately 5×10^{-5} M.

TABLE 3   Effect of proline concentration on mouth opening behavior in *P. tiarella*

Proline Concentration (M)	10^{-3}	10^{-4}	10^{-5}	10^{-6}
Number of Hydranths	23	24	21	21
Time for Mouth to Open (min)	1	2	4–5	>6
Average Time Mouth Remains Open (min)	25	14	9	2

Effect of Proline on Capitate Tentacles

The knob of the capitate tentacles appeared to shrink in volume when the hydranths were immersed in proline solutions. This response was readily visible after the hydranths were in 10^{-5} M proline for 3 to 4 minutes. During this shrinking, the cnidocils of the large stenoteles protruded prominently from the knob. Cnidocial protrusion does not seem to be a phase of that extrusion process by which the entire nematocyst's capsules are ejected undischarged; e.g., as seen when acontia eject their nematocysts in sodium citrate solution (Blanquet and Lenhoff 1966), because the capsules themselves do not protrude when the tentacles are in proline. Cnidocil protrusion in response to proline may simply be a nonspecific shrinking of the knob of the capitate tentacle in reaction to proline. Alternatively, it may represent a readiness response in which the nematocysts of the capitate tentacles are positioned for action by the proline emitted by the prey when it is first captured by the filiform tentacles. Because small molecules may initiate an array of feeding responses in hydroids (Lenhoff 1968), this alternative explanation does not seem unreasonable.

GENERAL DISCUSSION

The interrelationship between nematocysts and feeding response in cnidarians has been suggested by Loomis (1955) and by Lenhoff and Schneiderman (1959). They postulated that puncture of the prey by nematocysts leads to a leakage or bleeding of body fluids. These body fluids contain chemicals (e.g. proline, reduced glutathione) which, upon contact with the appropriate cnidarian receptor sites, initiate a feeding response.

In the case of the colonial gymnoblastic hydroid *Pennaria tiarella*, I have demonstrated some of the roles played by the filiform and capitate tentacles (with their respective nematocysts) in capturing and immobilizing prey. I have also shown that proline activates a feeding response which consists of the bending of the oral cone and the opening of the mouth. Apparently a gradient of proline emitted from the captured prey stimulates the oral cone to bend toward the prey. The open mouth, on contact with the prey, swallows it.

It would be tempting to suggest that the existence of proline-induced feeding responses in the gymnoblastic colonial hydroids *Cordylophora lacustris* and *Pennaria tiarella* are evidence for an evolutionary relationship between them. Because the specific activators of feeding are known for so few cnidarians, and because proline has recently been shown to be a feeding activator for such a distantly related cnidarian as the coral *Cyphastrea* (Mariscal, this vol.), to conclude an evolutionary relationship from these data alone is unwarranted.

SUMMARY

1. The role in food capture of the nematocysts on the filiform and capitate tentacles of *Pennaria tiarella* is discussed.

2. The feeding response of *Pennaria tiarella* to *Artemia* extracts consists of cone bending and mouth opening. A similar feeding response is induced by proline at concentration as low as 10^{-6} M. Pipecolic acid, a proline analog, also elicits a feeding response.

3. Proline causes the head of the capitate tentacles to shrink, thereby causing the cnidocils of the large stenoteles to be prominent.

LITERATURE CITED

Blanquet, R., and H. M. Lenhoff. 1966. A disulfide-linked collagenous protein of nematocyst capsules. *Science* **154**: 152–153.

Ewer, R. J. 1947. On the function and mode of action of the nematocysts of *Hydra*. *Proceedings, Zoological Society of London* **117**: 365–376.

Fulton, C. 1963. Proline control of the feeding reaction of *Cordylophora*. *J. General Physiology* **46**: 823–837.

Goreau, T. 1961. Comment. In *The biology of hydra and of some other coelenterates: 1961*, H. M. Lenhoff and W. F. Loomis, eds., p. 230. Coral Gables: University of Miami Press.

Josephson, R. K. 1961. Colonial response of hydroid polyps. *J. Experimental Biology* **38**: 559–577.

Lenhoff, H. M. 1968. Chemical perspectives on the feeding response, digestion, and nutrition of selected coelenterates. In *Chemical zoology,* vol. 2, M. Florkin and B. Scheer, eds., pp. 157–221. New York: Academic Press.

Lenhoff, H. M., and H. A. Schneiderman. 1959. The chemical control of feeding in the Portuguese man-of-war, *Physalia physalis* L. and its bearing on the evolution of the Cnidaria. *Biological Bulletin, Woods Hole* **116**: 452–460.

Loomis, W. F. 1955. Glutathione control of the specific feeding reactions of *Hydra*. *Annals of the New York Academy of Science* **62**: 209–228.

Lowry, O. H., N. Rosebrough, A. Farr, and R. Randall. 1951. Protein measurement with the Folin phenol reagent. *J. Biological Chemistry* **193**: 265–275.

K. JUNE LINDSTEDT CHAPTER 9
University of Southern California, Los Angeles

Valine Activation of Feeding in the Sea Anemone *Boloceroides*

Specific feeding activator compounds have been identified in relatively few coelenterates, though many coelenterates are known to respond to fresh tissue juices extracted from prey organisms (for a review, see Lenhoff 1968). The most extensive work in identifying specific feeding activators has been done with hydrozoans. Loomis (1955), using *Hydra littoralis,* demonstrated that its feeding response is controlled by the reduced tripeptide glutathione. Glutathione also controls the feeding response in the hydroid *Campanularia flexuosa* and in the siphonophore *Physalia physalis* (Lenhoff and Schneiderman 1959). Fulton (1963) found that the imino acid proline was the specific feeding activator in the colonial gymnoblastic hydroid *Cordylophora lacustris.* Proline was also shown by Pardy to be the feeding activator of another colonial gymnoblast *Pennaria tiarella* (this vol.).

In this chapter I present evidence showing chemical control of feeding in an anthozoan coelenterate; the feeding response of the Hawaiian swimming actinian, *Boloceroides* sp., is controlled by the branched amino acid valine.

MATERIALS AND METHODS

Maintenance of Organisms

Specimens of *Boloceroides* sp. were collected from reefs at the northern end of Coconut Island, Kaneohe Bay. Each animal was kept in 40 milliliters of artificial seawater in a covered Stender dish. This medium was "Instant Ocean," made up without adding the liquid trace elements provided by the manufacturer (Aquarium Systems Inc., Wickliffe, Ohio). The dishes were kept at least 18 inches from a Westinghouse 30-watt cool white fluorescent lamp. The medium was changed twice daily, and the animals were fed to repletion with *Artemia* nauplii every other day. All

---

An earlier version of this chapter appeared as "Valine activation of feeding in the sea anemone, *Boloceroides*" by K. June Lindstedt, L. Muscatine, and H.M. Lenhoff, in *Comparative Biochemistry and Physiology* **26**: 567–572.

experiments were performed on animals that had not been fed during the previous 24 hours. The size of the animals used, as measured by crown diameter (including tentacles), was approximately 1–3 centimeters. The researcher should be cautioned, however, that *Boloceroides* is capable of extreme changes in volume, particularly in the tentacles. Such volume changes are not unusual in healthy animals.

Feeding Response to Live Food

The feeding response of *Boloceroides* to live *Artemia* nauplii is well coordinated and distinctive. In animals that are not feeding, the tentacles are usually outstretched and appear relaxed (Figure 1). On contact with *Artemia* nauplii the tentacles elevate, sweep toward the central axis of the body, and begin to writhe (Figure 2). When a nauplius touches a tentacle, it is pierced by nematocysts and held. The tentacle is then brought rapidly toward the mouth of the anemone. The captured nauplius may be moved by ciliary action from the tentacle into the mouth, or the tentacle itself may enter the mouth with the *Artemia*.

Feeding Response to Homogenates and Extracts of Live Food

A similar feeding behavior could also be induced by a suspension of an *Artemia* homogenate. Dilutions of particle-free ethanol extract of that homogenate, however, did not evoke an active response. Therefore, in order to determine the exact nature of the feeding activator(s) in the extract, an assay had to be devised that would allow the use of soluble material, and that would take advantage of a reliable behavioral response of the anemone. Tentacle writhing alone could not be used as a behavioral assay for feeding activation because writhing can be induced by other stimuli. In fact, *Boloceroides* differs from most other solitary anemones because this anemone, through vigorous tentacle movements, can actually leave the substratum and swim (Josephson and March 1966).

Assay for a Feeding Response

The assay devised consisted of placing a 1-mm^2 piece of Whatman filter paper (no. 1 or 4) on the animal's mouth. A successful feeding response was scored if an anemone ingested the paper and retained it in the coelenteron for at least 5 minutes.

Before a test paper was presented to an animal, a control piece of untreated paper was first placed on the mouth. If the control paper was moved off the mouth within 2 minutes through ciliary action, then the animal was used in the experiment. Animals that ingested control paper were not used because of the possible presence of trace contamination of feeding activator. The 5-minute time was used because paper retained in the coelenteron for that length of time was always found to be retained for at least 1 hour or longer. If the anemone did reject the control paper, it did so within 2 minutes after the paper was placed on its mouth.

FIG. 1. *Boloceroides,* in the absence of a feeding activator.

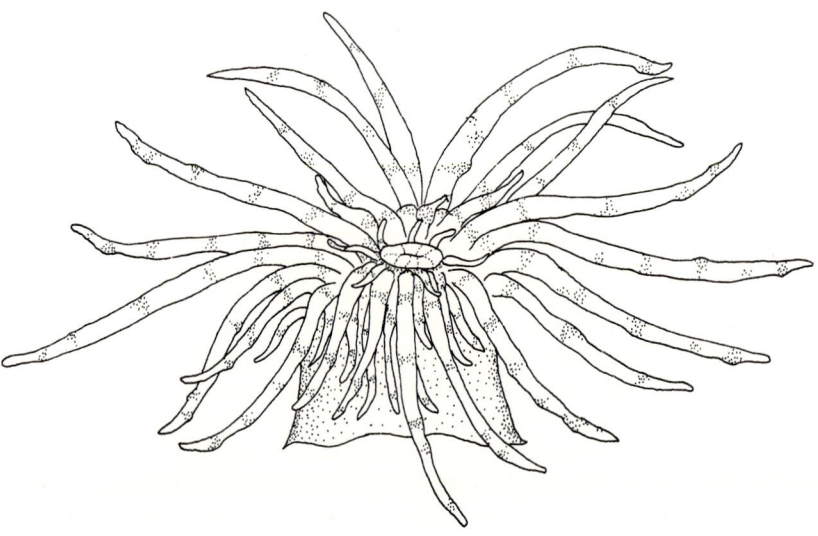

FIG. 2. *Boloceroides,* during a feeding response.

RESULTS

Effect of *Artemia* Extract Fractionated by Chromatography

Since preliminary study showed the feeding activator from *Artemia* nauplii to be unaffected by boiling or extraction with ethanol, the ethanol extract was further fractionated by chromatography. To prepare the extract, nauplii were washed in distilled water, dried, and homogenized in absolute ethanol. Following extraction, the suspension was centrifuged at 18,000 X g. Pieces of filter paper that had been immersed in the alcoholic supernatant and dried were immediately ingested and retained by *Boloceroides*.

The extract was spotted on Whatman no. 4 paper, and one-dimensional descending chromatograms were run in butanol-propionic acid-water (4 : 2 : 3) solvents (Bassham and Calvin 1957). The chromatograms were dried, and 1-mm^2 pieces of paper removed from different areas along the strip were placed on the mouths of anemones fasted for 1 day. The results showed that the best feeding response was elicited using pieces of paper removed from a position on the chromatogram corresponding to compounds with an R_F of 0.85.

To determine if the activator was an amino acid having an R_F approximating the observed one, I offered anemones bits of paper that previously had been treated with one application of 1 M solution of suspect (and related) amino acids. The results, summarized in Table 1, show that all the amino acids (Calbiochem, Los Angeles), as well as unspotted paper, produced some positive responses, but that the most consistent response was produced by the branched amino acid valine.

I discounted from consideration responses under 30 percent because the control experiments with clean filter paper showed that 27 percent (3 out of 11) of the animals that had appeared acceptable for

TABLE 1 Feeding responses of *Boloceroides* sp. to amino acids on filter paper

Amino Acid	Feeding Response by Number of Animals		Positive Responses
	(+)	(-)	(%)
None	3	8	27
Tyrosine	4	16	20
Methionine	6	16	27
Proline	1	10	9
Glutamate	2	9	18
Leucine	7	20	26
Isoleucine	11	13	46
Valine	50	5	91

experimentation (i.e., they initially rejected clean filter paper) would ingest a second piece of clean paper immediately afterward. I can offer no explanation for this "background" feeding behavior to apparently chemically inert material.

Assay of Valine and of Isoleucine

Because significant responses were also given to the branched amino acid isoleucine, experiments were carried out comparing the time required for *Boloceroides* sp. to ingest filter paper treated with either valine or isoleucine. The results showed that 30 out of 32 animals ingested the valine-treated paper within 1 minute, whereas only 7 out of 20 anemones ingested the isoleucine-treated paper in this time. The response to valine by the anemones was always faster than the response to isoleucine, and was frequently accompanied by active tentacle twitchings, elevations, and writhings. Never did such tentacle activity occur in response to isoleucine.

The anemones would also ingest clean filter paper if they were in seawater containing dissolved valine. Best results were obtained in 10^{-3} M to 10^{-5} M valine, with 8 to 10 of each group of 10 animals responding. At 10^{-6} only 5 of the 10 animals in the group gave positive responses.

I also observed that, in the absence of bits of filter paper, valine solutions or *Artemia* extracts repeatedly caused a noticeable swelling of the mouth area. This swelling, however, occasionally occurred on mechanical stimulation of the anemone.

Further proof that valine was the feeding activator in the *Artemia* extracts was obtained by chromatographing a solution of the extract to which had been added some ^{14}C-valine that had a high specific activity. A radioautograph was made of the chromatogram using Kodak blue sensitive medical X-ray film. Pieces cut from the spot on the paper having radioactivity were presented to several anemones and were ingested in each instance.

Although the test paper used in these experiments was applied directly to the mouth, valine receptor sites are probably also located on the tentacles, because anemones ingested valine-treated paper placed there.

Effects of Leucine and of Isoleucine on the Valine-activated Response

The chemical receptor was shown to be specific for the size and shape of valine through experiments testing leucine and isoleucine as possible competitive inhibitors. The experiments consisted of placing untreated or valine-treated paper on the oral disc of an anemone while either leucine or isoleucine at 10^{-4} M was in the medium. The results (Table 2) show that leucine at 10^{-4} M exerted no appreciable inhibition on the ingestion of the valine-treated paper. On the other hand, the same

TABLE 2 Effects of leucine and isoleucine on the valine-stimulated feeding response of *Boloceroides* sp.

Amino Acid in Medium (10^{-4} M)	Amino Acid on Filter Paper	Feeding Response by Number of Animals		Positive Responses
		(+)	(−)	(%)
Leucine	None	3	14	17
	Valine	14	4	78
Isoleucine	None	0	9	0
	Valine	5	11	31
None	Valine	10	0	100

concentration of isoleucine inhibited two-thirds of the animals from ingesting the valine-treated paper.

These analog experiments point out that the structure of the valine molecule necessary to activate the *Boloceroides* receptor is the α-amino-*n*-butyric acid with an ethyl group at the branch point, i.e., isoleucine, is not sufficiently specific to give a strong feeding response (Table 1), but that it does have enough structural similarity to valine to be an effective competitive inhibitor (Table 2).

That leucine is unable to inhibit competitively the ingestion of valine-treated paper supports my view of the active structure of valine needed to induce a response. Leucine is identical to valine in all respects except that the branch point is separated from the α-carbon by an additional methylene group. This slight lengthening of the backbone of the amino acid accounts for the ineffectiveness of leucine as either an activator or an inhibitor. It is interesting to compare this finding with Loomis's discovery (1955) that asparthione, which is one methylene group shorter than glutathione in the γ-glutamyl moiety, does not elicit a feeding response in hydra.

DISCUSSION

In addition to its unique specificity for valine, the feeding response of *Boloceroides* differs from that described for other coelenterates in which a specific activator has been identified. Solutions of glutathione cause the tentacles of *Hydra littoralis* to bend and the mouth to open (Loomis 1955; Lenhoff 1961). Mouth opening is also a significant feature of the proline activation of *Cordylophora* (Fulton 1963), *Pennaria* (Pardy, this vol.) and *Cyphastrea* (Mariscal, this vol.). In *Boloceroides,* however, no visible mouth opening occurs with valine in solution. Furthermore, it is difficult to detect unequivocally a feeding response in *Boloceroides* unless a solid particle is present on the animal. Hence, it appears that some sort of mechanical stimulation is needed along with the chemical activation. Such a combined action of chemical and particle suggests that the cilia along the surface of the animal are influenced by the feeding activator, as pointed out in other actinians by G. H. Parker many years ago (1896).

Undoubtedly, other coelenterates give a distinctive feeding response to a specific chemical activator only in the presence of mechanical stimuli. In fact, in those reported instances of coelenterates ingesting so-called solid objects in the absence of an added chemical stimulator, as recently described, for example, with *Epiactis prolifera* (Lenhoff 1965), there remains the strong possibility that traces of chemical activators are present on the object or in the water bathing the animals.

From the results of the present study, we can now add valine to the growing list of small molecules known to coordinate aspects of the feeding movements of some coelenterates. Of these compounds—glutathione (Loomis 1955; Lenhoff 1961), proline (Fulton 1963; Pardy, this vol.; Mariscal, this vol.), valine and tyrosine (Blanquet and Lenhoff 1968)—valine is the first one shown to induce a feeding response in an actinian. Recently, using an unidentified small acontiate sea anemone found floating on a sargassum weed in Biscayne Bay, Miami, Florida, N. Smith and H. M. Lenhoff tentatively identified glutamine as a feeding activator. Whether the association of specific feeding activators with specific groups of coelenterates is of taxonomic or evolutionary significance, or whether they are strictly random associations, might be better discussed after a greater number of coelenterates have been investigated.

SUMMARY

1. The feeding response of the swimming sea anemone *Boloceroides* sp. was found to be activated by the amino acid valine.

2. Isoleucine served as a competitive inhibitor, whereas leucine did not.

3. No feeding response to valine dissolved in seawater was observed unless the animals were presented at the same time with an inert object.

LITERATURE CITED

Bassham, J. A., and M. Calvin 1957. The path of carbon in photosynthesis, Englewood Cliffs, N.J.: Prentice-Hall.

Blanquet, R., and H. M. Lenhoff. 1968. Tyrosine enteroreceptor in hydra: Its function in initiating a new behavioral modification. *Science* 159: 633–634.

Fulton, C. 1963. Proline control of the feeding reaction of *Cordylophora*. *J. General Physiology* 46: 823–837.

Josephson, R. K., and S. C. March. 1966. The swimming performance of the sea-anemone *Boloceroides*. *J. Experimental Biology* **44**: 493–506.

Lenhoff, H. M. 1961. Activation of the feeding reflex in *Hydra littoralis*. I. Role played by reduced glutathione, and quantitative assay of the feeding reflex. *J. General Physiology* **45**: 331–344.

———. 1965. Mechanical stimulation of feeding in *Epiactis prolifera. Nature*. London **207**: 1003.

———. 1968. Chemical perspectives on the feeding response, digestion, and nutrition of selected coelenterates. In *Chemical zoology*, vol. 2, M. Florkin and B. Scheer, eds., pp. 157–221. New York: Academic Press.

Lenhoff, H. M., and H. A. Schneiderman. 1959. The chemical control of feeding in the Portuguese man-of-war, *Physalia physalis* L. and its bearing on the evolution of the Cnidaria. *Biological Bulletin, Woods Hole* **116**: 452–460.

Loomis, W. F. 1955. Glutathione control of specific feeding reactions of hydra. *Annals, New York Academy of Sciences* **62**: 209–228.

Parker, G. H. 1896. The reactions of *Metridium* to food and other substances. *Bulletin, Museum of Comparative Zoology at Harvard College* **29**: 107–119.

RICHARD N. MARISCAL
Florida State University, Tallahassee

CHAPTER 10

The Chemical Control of the Feeding Behavior in Some Hawaiian Corals

Since Loomis (1955) first demonstrated that the reduced tripeptide glutathione elicited feeding behavior in *Hydra littoralis*, a number of workers have demonstrated a similar phenomenon among other coelenterates. Lenhoff and Schneiderman (1959) found that the siphonophore *Physalia physalis* and the calyptoblastic hydroid *Campanularia flexuosa*, also gave feeding responses to glutathione. On the other hand, Fulton (1963) showed that the feeding behavior of the gymnoblastic brackish-water colonial hydroid *Cordylophora lacustris* is elicited not by glutathione, but by the imino acid proline. Pardy (this vol.) found that proline causes mouth opening in the gymnoblastic marine hydroid *Pennaria tiarella*. Lindstedt (this vol.) found that the amino acid valine initiated feeding in the swimming sea anemone *Boloceroides*. Smith and Lenhoff (unpublished) showed that glutamine is a feeding stimulator in an unidentified acontiate sea anemone. For recent reviews on the chemical mediators of feeding and other behavior in coelenterates, see Lenhoff (1967, 1968).

Nothing is known of the specific feeding stimuli for corals except that mouth opening and feeding behavior can be elicited by tissue extracts (Abe 1938). Therefore, we conducted a comparative study of the stimuli for mouth opening in three Hawaiian corals.

Most of the work involved the colonial encrusting coral *Cyphastrea ocellina*. Both proline and, to a lesser extent, reduced glutathione were effective in causing mouth opening in this species. Analogs of these compounds, pipecolic acid and S-methyl glutathione, respectively, also caused mouth opening in *Cyphastrea*.

An earlier version of this chapter appeared as "The chemical control of feeding behavior in *Cyphastrea ocellina* and in some other Hawaiian corals" by Richard N. Mariscal and Howard M. Lenhoff, in *Journal of Experimental Biology* 49 (3): 689–699.

MATERIALS AND METHODS

The corals were collected from the reefs in Kaneohe Bay and were maintained in running-seawater tables. They were kept without additional food for 24 hours before use, and were used only once in any 24-hour period. Because the animals gave similar feeding behavior in either artificial seawater or natural seawater, the latter was used almost exclusively.

The pH of the various experimental solutions was similar to that of the natural seawater in which they were maintained; therefore, no additional buffer was added. The temperature of the solutions ranged between 26°–28° C.

Aqueous homogenates and extracts of either the nauplii of the brine shrimp *Artemia salina* or of freshly collected plankton were prepared as follows: solid-pack suspensions of a known volume of *Artemia* nauplii were homogenized in a small volume of seawater. The plankton (consisting mostly of crab megalops, mysids, caridean larvae, and chaetognaths) was also homogenized in a small volume of seawater. The homogenates were either used directly, or were centrifuged and only the extracts tested. Since no differences in feeding responses could be detected using either source of extracts, the *Artemia* were primarily used in the chromatographic analyses.

Extracts of *Artemia* or plankton for chromatographic analyses were obtained by homogenizing the animals in distilled water, centrifuging the homogenate at about $4,500 \times g$, and mixing the resultant aqueous supernatant with an equal volume of 70 percent ethyl alcohol. The alcohol-soluble material was dried on a Buchler Rotary Evapo-Mix (Buchler Instruments, New York, N. Y.) and then resuspended in 70 percent ethyl alcohol. This procedure was repeated several times, first with 70 percent ethyl alcohol, then with 95 percent ethyl alcohol, and finally with absolute alcohol, until no further precipitate was formed.

Different concentrations of the alcohol extracts were streaked on Whatman no. 4 filter paper. One-dimensional descending paper chromatograms were prepared using a butanol-propionic acid solvent after the technique of Benson et al. (1950). One strip was developed with 0.25 M ninhydrin in acetone and its cochromatogram was retained for testing. When a strip which gave both good concentration and separation of spots was found, pieces were cut of the undeveloped portion of the cochromatogram and presented directly to the coral polyps. In this way the approximate region of the chromatogram which elicited mouth opening could be determined. A piece of blank filter paper which had also been run through the solvent system served as a control.

To determine the identity of the compounds present on the

chromatograms of *Artemia* extract which elicited a feeding response, the R_F of each active spot was compared with those of known amino acids that had been run through the same one-dimensional chromatography system.

Once an active area on the paper was found, small pieces of clean filter paper were spotted with known concentrations of various amino acids having similar R_F values. When one of these pieces of treated filter paper elicited a response, a solution of known molar concentration of the same compound was prepared, and the response of the entire coral colony to this compound observed and recorded.

The commercially available amino acids which caused positive responses were then separated by chromatography to determine if any impurities were present. Pieces of paper obtained from the unsprayed cochromatograms were tested on the corals to ensure that the compound in question was responsible for the observed response.

All microscopic observations were made with a Bausch and Lomb Stereozoom dissecting microscope.

The technique of Lenhoff (1961) for quantifying the mouth-opening response of *Cyphastrea* proved to be impractical because of the extremely long time the coral polyps' mouths remained open in solution. Three alternative methods were employed.

Method I was a simple plus-minus system which indicated the extent of mouth opening of *Cyphastrea*. Three pluses signified the maximum mouth opening (about 0.5 mm) of the individual polyps, whereas a minus meant no visible response. Method II involved counting the number of polyps with open mouths in the experimental solution after a fixed interval of time. A 15-minute interval proved convenient for *Cyphastrea* since this allowed a near-maximal response of the coral colony in the highest concentration of compound used (10^{-3} M). Lesser concentrations of compounds invariably caused a lesser response after the same time interval. Method III involved placing the coral in the experimental solution and then counting the number of *Cyphastrea* polyps out of 50 responding with time.

RESULTS

Feeding Responses in *Cyphastrea* to Live Prey, and to Homogenates or Extracts of Live Prey

Feeding behavior in *Cyphastrea* (as well as in the other corals investigated) in response to live prey (plankters or *Artemia*), homogenates, or seawater extracts of prey, consisted of a wide mouth opening (Figures 1, 2). Following exposure to any of these stimuli, *Cyphastrea* responded by giving a short, sharp contraction of the tentacles and/or oral-disc area. Within 1–3 seconds the mouths of some of the

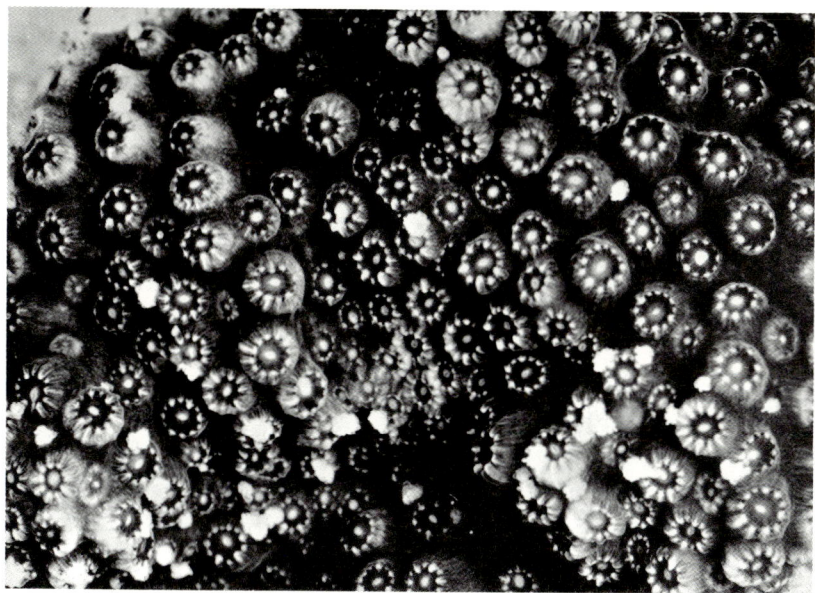

FIG. 1. Colony of *Cyphastrea ocellina* in 10^{-3} *M* reduced glutathione prepared in natural seawater. Note the open mouths of the polyps and the extruded mesenterial filaments (white puffs) on the surface of the colony.

polyps opened, gradually followed by others, until after a period of time, dependent upon the concentration of homogenate or extract or upon the number of prey organisms, nearly all the polyps had responded. Given enough time (30 to 60 minutes), all the polyps of a colony would generally respond to the aforementioned stimuli. While the mouth was opening, the prey organisms or homogenate particles were worked toward and into it by either ciliary action, muscular action, or both.

Dilutions as low as 1- to 10-million (10^{-7}) parts of the homogenate from a solid pack of *Artemia* nauplii still gave a definite feeding response as measured by Method III (Figure 3). There was, however, a decided difference in the number of polyps responding with time as well as in the extent to which the mouth opened in the various dilutions of homogenate or extract (Table 1).

Identification of the Compounds Which Stimulate Feeding in *Cyphastrea*

Preliminary tests revealed that the stimulus contained in *Artemia* aqueous extract which caused mouth opening of *Cyphastrea* was unaffected by boiling and soluble in 95 percent ethyl alcohol. Eluates of chromatogrammed *Artemia* alcohol extracts tested on *Cyphastrea* showed

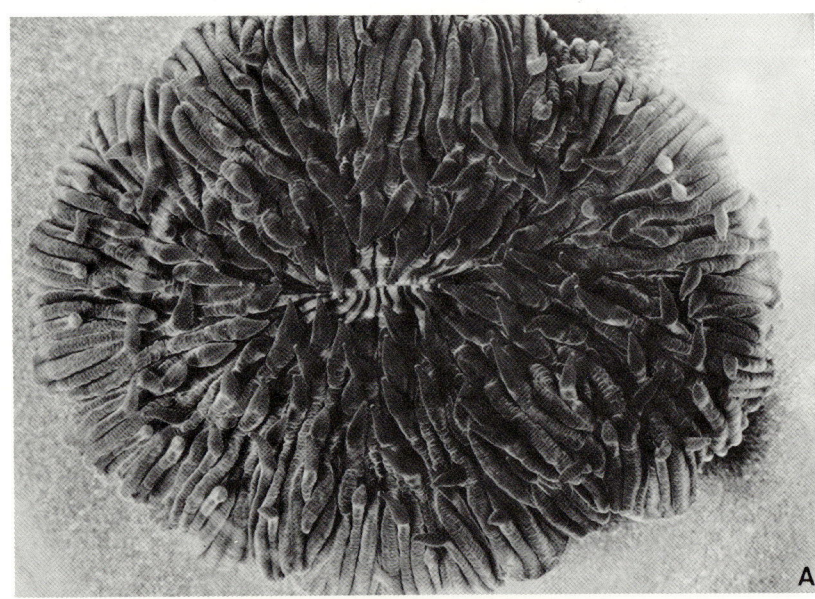

FIG. 2. A, *Fungia Scutaria* polyp in natural seawater in the absence of any feeding activator. B, Same *Fungia scutaria* polyps in 10^{-5} M proline prepared in natural seawater.

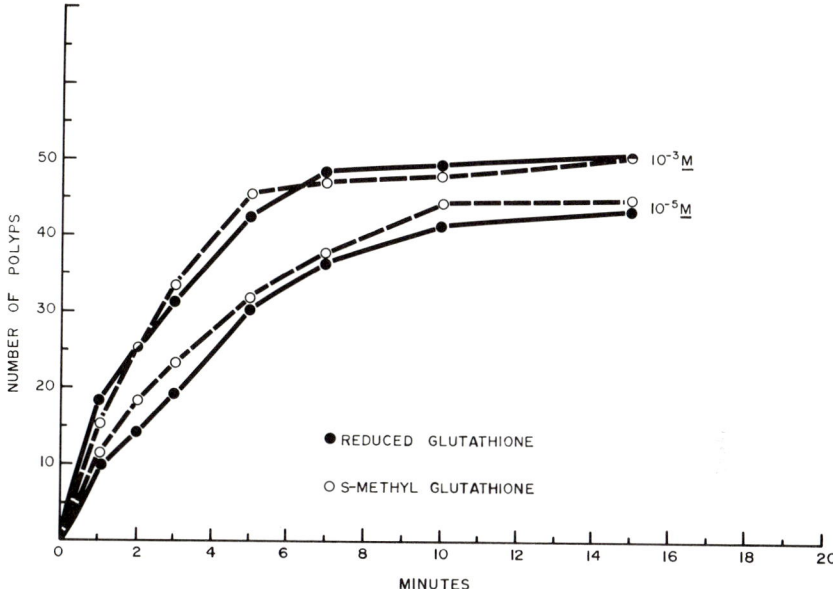

FIG. 3. The effect of various dilutions of *Artemia* homogenate on the numbers of *Cyphastrea ocellina* polyps giving a feeding response.

that the major activity for mouth opening was restricted to a spot having an R_F value of 0.6. Somewhat less activity appeared lower on the chromatographic strip (R_F about 0.3).

Because the upper spot had the yellow ninhydrin color characteristic of proline as well as having a similar R_F value, several tests were conducted with this amino acid. A piece of filter paper treated directly with a solution of 10^{-1} M proline and a similarly treated piece which had been through the butanol-propionic acid solvent system were both presented to individual *Cyphastrea* polyps. In all the experiments conducted using both of the above types of proline-treated filter paper, a wide mouth opening was elicited. Other amino acids each having an R_F value close to that of proline and similarly applied to filter paper provoked no response when presented to *Cyphastrea* polyps.

Since reduced glutathione was thought to have an R_F value similar to that of the lower active spot, pieces of filter paper treated as above for proline were presented to *Cyphastrea* polyps. A definite mouth-opening response, although not quite as strong as that observed with proline, was immediately produced. Filter paper treated with the various amino acids having R_F values corresponding to that of the lower active spot did not cause any distinct mouth-opening response. Reduced glutathione has

TABLE 1 Mouth-opening behavior of *Cyphastrea ocellina* in various solutions

(M)	Proline	Pipecolic Acid	Hydroxyproline	Reduced Glutathione	S-methyl Glutathione	*Artemia* Homogenate*
10^{-2}	0	0	0	0	0	+++
10^{-3}	+++	+++	++	++	++	+++
10^{-4}	0	0	0	0	0	++
10^{-5}	++	++	0	+	+	++
10^{-6}	0	0	0	0	0	++
10^{-7}	+	+	0	–	–	+

Key: +++, wide mouth opening; ++, moderate mouth opening; +, slight mouth opening; –, no mouth opening; 0, no data.
*Concentrations of *Artemia* extract refer to dilutions of the homogenized packed nauplii.

TABLE 2 Mouth-opening behavior of *Cyphastrea ocellina* after 15 minutes in various solutions

	Proline (M)			Hydroxy-proline (M)	Pipecolic Acid (M)			Reduced Glutathione (M)			S-methyl Glutathione (M)	
	10^{-3}	10^{-5}	10^{-7}	10^{-3}	10^{-3}	10^{-5}	10^{-7}	10^{-3}	10^{-5}	10^{-7}	10^{-3}	10^{-5}
Number of Polyps	562	979	237	173	245	91	126	894	615	222	290	273
Mouths Open	515	842	174	153	245	87	115	818	543	17	281	215
Percentage Responding	91.6	86.0	73.4	88.4	100	95.6	91.3	91.5	88.8	7.7	96.9	78.8

recently been shown to be abundant in *Artemia* nauplii by R. D. Brown (personal communication) using the fluorometric assay of Cohn and Lyle (1966).

Responses of *Cyphastrea* to Analogs of Proline and Reduced Glutathione

Pipecolic acid. Pipecolic acid, a six-membered ring analog of proline, has been found to date only in plant material. It proved to be as effective (and in some cases more effective) as proline in causing mouth opening in *Cyphastrea*. With Method I, the same concentrations of proline and pipecolic acid elicited similar degrees of mouth opening (Table 1).

Using Method II, it was found that pipecolic acid caused a greater percentage of response at all three concentrations used than did proline (Table 2). Fulton (1963) reported that the response of the brackish-water hydroid *Cordylophora* in pipecolic acid was approximately one-tenth that in proline. Pardy (this vol.) found that pipecolic acid was also one-tenth as effective as proline in causing mouth opening in the marine hydroid *Pennaria tiarella*.

When the number of *Cyphastrea* polyps out of 50 responding with time were plotted (Method III), a striking difference in the shape of the curves obtained with proline and pipecolic acid was observed (Figure 4).

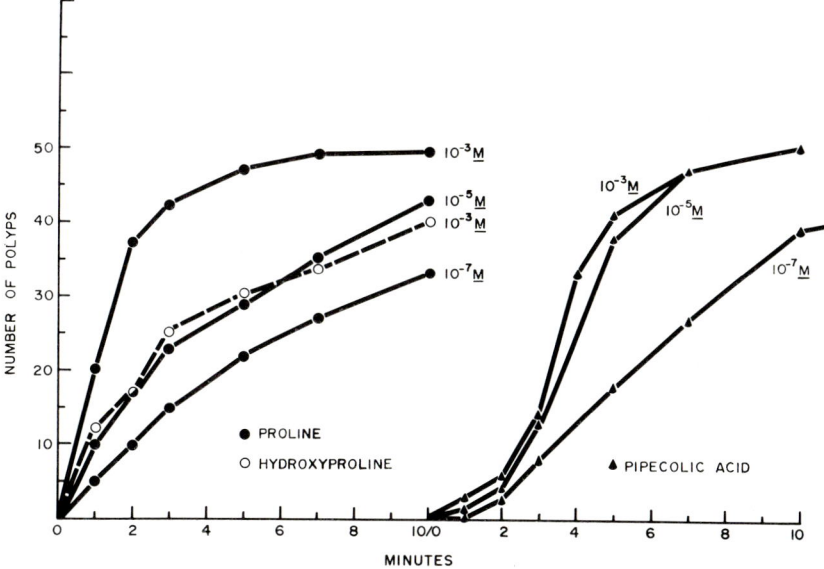

FIG. 4. The effect of various concentrations of proline, hydroxyproline, and pipecolic acid on the numbers of *Cyphastrea ocellina* polyps giving a feeding response.

The curve representing the increase in number of polyps responding with time to proline was hyperbolic, while the curves obtained at all three concentrations of pipecolic acid tested were distinctly sigmoid, with 1-2-minute lag periods. Such a lag period was never observed with proline or with the other compounds tested.

Hydroxyproline. Hydroxyproline, which can be considered a proline analog, was found to be one one-hundredth as effective as proline in tests with *Cyphastrea*. In the various tests conducted, a concentration of 10^{-3} M hydroxyproline elicited a response very similar to that from 10^{-5} M proline (Tables 1, 2; Figure 4). As Fulton (1963) warned, however, commercially available hydroxyproline may have a 1 percent contamination by proline and this could account for the response of *Cyphastrea* to solutions of hydroxyproline.

S-methyl glutathione. S-methyl glutathione is a rare analog of reduced glutathione which has only recently been described as occurring naturally in the bovine brain (Kanazawa et al. 1965). Synthetic S-methyl glutathione (Zion Chemical Company, Yavne, Israel) proved to be as effective in eliciting a mouth-opening response in *Cyphastrea* as corresponding concentrations of reduced glutathione (Tables 1, 2; Figure 5).

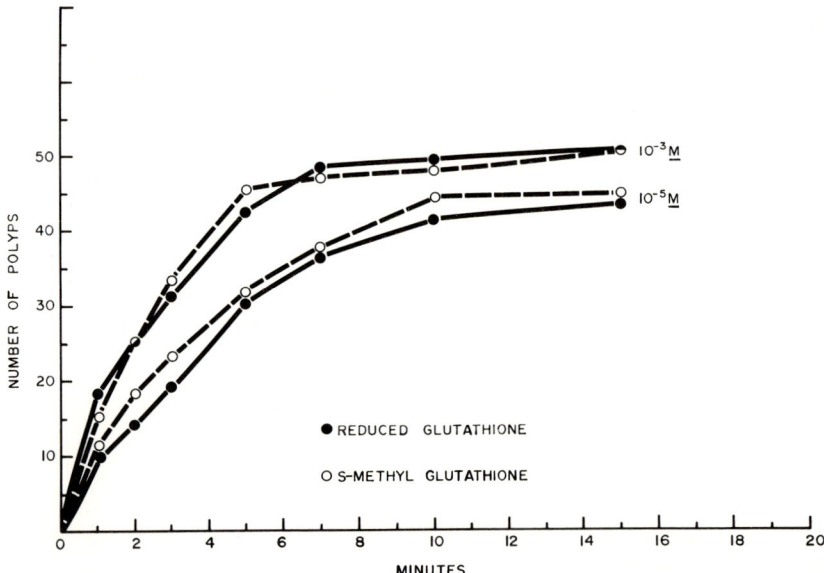

FIG. 5. The effect of various concentrations of reduced glutathione and S-methyl glutathione on the numbers of *Cyphastrea ocellina* polyps giving a feeding response.

The extent of mouth opening in *Cyphastrea* in reduced glutathione (and S-methyl glutathione) was one order of magnitude less than in the corresponding concentrations of proline or pipecolic acid (Table 1). Although the percentage of mouths opening (Table 2) for the two groups of compounds at 10^{-3} M and 10^{-5} M was roughly similar, both proline and pipecolic acid evoked much stronger responses at 10^{-7} M than did reduced glutathione. Thus proline seems to be more effective than reduced glutathione in causing *Cyphastrea* polyps to open their mouths.

Response of *Cyphastrea* to Progressively Higher Concentrations of Glutathione

Cyphastrea polyps showed greater responses to glutathione as the concentration increased (Table 2). Hence, it was postulated that if polyps were transferred from a lower glutathione concentration to a higher one, they would give roughly the same response as they would have given had they been initially introduced into the higher concentration. This proved to be the case. Whereas 10.9 percent of the polyps tested responded to 10^{-7} M glutathione, an 80.8 percent response was evoked by the same polyps when transferred to 10^{-5} M glutathione (Table 3).

Extrusion of Mesenterial Filaments

Whereas proline was the primary stimulus for mouth opening in *Cyphastrea*, glutathione appeared to be more effective in causing mesenterial filament extrusion. Although the mouths of the polyps in 10^{-3} M glutathione did not open as widely as those in 10^{-3} M proline, there were many more mesenterial filaments extruded on the surface of the colony in glutathione (Figure 1). We could not find, however, a strong and consistent correlation of the number of filaments extruded with the molarity of the solution. Inasmuch as this response was not studied in detail, a good deal more work will be necessary to determine, not only the relationship of mesenterial filament extrusion to prey capture and feeding in general, but also the influence of the various chemical compounds on the filament extrusion process itself.

TABLE 3 Mouth-opening behavior of *Cyphastrea ocellina* colonies after transfer from lower to higher concentrations of reduced glutathione

	Reduced Glutathione (M)		
	10^{-3}	10^{-5}	10^{-7}
Number of Polyps	334	89	220
Mouths Open	321	72	24
Percentage Responding	96.1	80.8	10.9

Comparative Survey of the Feeding Response in Corals Other Than *Cyphastrea*

Fungia scutaria. Mucus secreted by *Fungia* may play a greater role in feeding than that secreted by *Cyphastrea*. When small crustaceans contact the oral disc of *Fungia*, they appear to be trapped in the mucus itself as well as being immobilized by the nematocysts. The mucous sheet with its trapped organisms is then carried to the mouth by ciliary action (see Abe 1938).

Within 30–60 seconds of prey capture or the introduction of a food substance, *Fungia* responds with a wide mouth opening (Figure 2). Depending on the concentration of the stimulatory substance, the mouth may remain open for several hours. In addition, the tentacles bend toward the mouth in concurrence with a general swelling or "inflating" of the tissue immediately surrounding the mouth. The lips of the mouth often bend toward a localized stimulus, such as a small piece of food on the disc.

Small portions of filter paper containing chromatogrammed alcohol extracts of *Artemia* were presented to *Fungia* as well as to *Cyphastrea*. Most mouth-opening activity was caused by spots having R_F values between 0.58 and 0.66. These experiments were not always repeatable, however. Further experiments similar to those conducted with *Cyphastrea* revealed that proline appeared to be one inducer for mouth opening. In addition, methionine, tyrosine, and reduced glutathione caused mouth opening occasionally. It is possible that other compounds may prove to be effective in the case of *Fungia*, and the above list should not be considered exhaustive.

In the initial experiments with *Fungia*, small (3–6 cm diameter) isolated polyps were used. However, a good deal of individual variation in response was evident when small numbers of animals were used. This variability was present whether the animals were kept in artificial or natural seawater, or whether they were fed or starved for prescribed periods prior to use. Because relatively large numbers of small stalked *Fungia* were readily available, these were used to provide statistically more reliable data in subsequent experiments. The response of *Fungia* to proline and reduced glutathione in solution was tested for comparison with similar data from *Cyphastrea*. In spite of the variability in response previously observed with *Fungia*, these experiments using larger numbers of animals gave good reproducibility (Table 4).

Note in Table 4 that 10^{-3} M proline caused the highest percentage of animals to open their mouths (94.2 percent). At this concentration the mouths occasionally remained open for several hours. A concentration of 10^{-3} M glutathione elicited only a 70.4 percent response, and the mouths stayed open about 12 minutes maximum.

TABLE 4 Mouth-opening behavior of *Fungia scutaria* after 2 minutes in test solution

	Proline (M)			Pipecolic Acid (M)		Reduced Glutathione (M)			S-methyl Glutathione (M)
	10^{-3}	10^{-5}	10^{-7}	10^{-3}	10^{-5}	10^{-3}	10^{-5}	10^{-7}	10^{-5}
Number of Polyps	103	103	103	103	103	196	194	103	103
Mouths Open	97	42	41	86	51	138	95	33	33
Percentage Responding	94.2	40.8	39.8	83.4	49.5	70.4	48.9	32.0	32.0

Pocillopora damicornis. A feeding stimulus caused a sharp contraction and partial withdrawal of individual *Pocillopora* polyps, followed by a wide mouth opening similar to that of *Cyphastrea*. Both the contraction and mouth opening seemed important for the full feeding response.

Pieces of a filter-paper chromatogram of *Artemia* alcoholic extract gave results similar to those obtained with *Cyphastrea*. Because of the difficulties in observing these small polyps, only a few experiments were conducted with *Pocillopora damicornis*. These revealed that proline and reduced glutathione seemed to be the main stimuli for mouth-opening (Table 5). In addition, there were occasional responses to methionine and possibly phenylalanine. Note that the percentages of responses were similar in both 10^{-3} M proline and 10^{-3} M glutathione. Proline, however, elicited a stronger contraction and wider mouth-opening response than did a similar concentration of glutathione spotted on filter paper, which implies that proline was more effective than glutathione in the mouth-opening response of *Pocillopora*. In contrast to *Cyphastrea*, *Pocillopora* extruded more mesenterial filaments in proline than in glutathione.

Other corals. The feeding responses of four additional species of corals—*Tubastrea manni*, *Porites compressa*, *Pocillopora meandrina*, and *Leptastrea bottae*—were also studied (Table 6). For comparison, the results of similar experiments with *Cyphastrea ocellina*, *Fungia scutaria*, and *Pocillopora damicornis* are presented in the same table.

The best feeding responses were generally elicited by proline. Reduced glutathione was also effective, but less so than proline. However, in one coral, *Leptastrea bottae*, glutathione appeared to elicit a stronger mouth-opening response than did proline. Because only a few specimens of this species were available at the time, no further experiments were conducted.

The only coral which did not give good mouth-opening responses to either proline or reduced glutathione was *Porites compressa*. Only a few experiments were carried out with this species, however, and it remains uncertain what compounds stimulate its feeding.

TABLE 5 Mouth-opening behavior of *Pocillopora damicornis* after 15 minutes in test solution

	Proline (M)			Reduced Glutathione (M)	
	10^{-3}	10^{-5}	10^{-7}	10^{-3}	10^{-5}
Number of Polyps	311	240	204	86	98
Mouths Open	295	202	134	79	87
Percentage Responding	94.8	84.2	65.7	91.8	88.7

TABLE 6 Responses of corals to pieces of filter paper dipped in 10^{-1} M concentrations of various compounds

Compounds	Pocillopora damicornis	Cyphastrea ocellina	Tubastrea manni	Fungia scutaria	Porites compressa	Pocillopora meandrina	Leptastrea bottae
Proline	+++	+++	+++	+++	++	+++	++
Reduced Glutathione	++	++	++	++	++	+++	+++
Methionine	±	−	−	±			
Tyrosine	−	−	−	±			
Valine	−	−	−	−			
Tryptophan	−	−		−			
Leucine	−	−		−			
Phenylalanine	+	−		−			
	*						

Key: +++, wide mouth opening, strong tentacle contraction; ++, moderate mouth opening, tentacle contraction; +, slight mouth opening, no contraction; ±, no mouth opening, slight contraction; −, no mouth opening, no contraction; *, occasional slight mouth opening, no contraction.

DISCUSSION

The general pattern of prey capture in *Cyphastrea* can be summarized as follows: (*a*) a small crustacean such as a crab megalops either bumps into the tentacle of an expanded polyp or blunders into the skeletal cup of a partially contracted one. (*b*) the tentacles, sensing the contact, infold sharply, partially entrapping the prey as well as bringing it into close contact with the mouth of the polyp. Simultaneously, the contact with the tentacles causes the nematocysts there to discharge and puncture the prey, thereby releasing the crustacean's body fluids, which contain numerous compounds including proline and glutathione. (*c*) such chemical stimulators of feeding cause the mouth to begin to open. (*d*) often the side of the mouth nearest the prey begins to bend toward it. This orientation of the mouth to the prey cannot be evoked by strictly mechanical stimuli, such as contact with a small piece of clean filter paper. A similar size piece of filter paper soaked in crustacean extract, proline, or glutathione, however, always evoked a mouth-opening response like that caused by live prey.

Actual physical contact with the mouth is not necessary to cause mouth opening. Generally, the closer the captured prey to the mouth, the more rapid the mouth opening. As the mouth opens, ciliary currents as well as muscular action seem to direct the prey toward and into it. When the lips contact the prey, they appear to work their way up and around it as it is being drawn into the coelenteron.

This study presents the first instance in which the feeding behavior of a single coelenterate, *Cyphastrea*, has been shown to be controlled by either a specific amino acid or a specific peptide (as well as their respective analogs). To date, all coelenterates investigated have been shown to respond to glutathione alone (e.g., various species of hydra; *Physalia*) or to a specific amino acid (e.g., *Cordylophora, Pennaria, Boloceroides*), but not to both. Although additional compounds may be found to induce feeding in other corals, *Cyphastrea ocellina* seems to respond primarily to proline or reduced glutathione, with proline being the more effective activator at low concentrations.

The response curve of *Cyphastrea* to pipecolic acid (Figure 4) was unusual in that it was sigmoid, while those to proline, glutathione, and S-methyl glutathione were hyperbolic (Figures 4 and 5). The initial 1–2-minute lag period in responsiveness does not reflect a lowered sensitivity of *Cyphastrea* to pipecolic acid, because a greater percentage of polyps responded to this compound than to proline at all concentrations tested (Table 2). The sigmoid curve suggests that the polyps are responding to pipecolic acid in a facilitated fashion, perhaps nervous, throughout the colony. Alternatively, the pipecolic acid might be involved in an allosteric activation of the receptor sites as described by Monod, Wyman, and Changeux (1965).

As Fulton (1963) pointed out, although the proline and glutathione molecules are markedly different, some glutathione molecules in solution may take a form having a heterocyclic a-amino acid moiety (see Calvin 1954; Wieland 1954; Isherwood 1959). Possibly this form of glutathione might be recognized by the proline receptors of *Cyphastrea*. Because only a small proportion of the glutathione molecules might have an a-amino structure at any one time, we might expect higher concentrations of glutathione to be necessary to give a response as effective as that given by proline alone. Such is the case. Another possibility is that two different receptor sites are involved—one for proline and one for glutathione.

The number of mesenterial filaments extruded in various concentrations of proline and glutathione was checked for *Cyphastrea* and *Pocillopora damicornis*. *Cyphastrea* extruded a greater percentage of filaments in glutathione than in proline. *Pocillopora*, however, extruded more filaments in proline than in glutathione, and always extruded a relatively greater number than did *Cyphastrea* in either solution. In high concentrations of proline (10^{-3} M) *Pocillopora* extruded approximately one filament for each two polyps observed. Concentrations of 10^{-5} M and 10^{-7} M proline caused *Pocillopora* to extrude only about 10 percent of its filaments. With either species, however, there seemed to be no fixed correlation between the concentration of compounds which induced mouth opening and the number of mesenterial filaments extruded. Some filaments were extruded in feeding experiments with live crustaceans, but these did not seem essential for prey capture. It is possible that the mesenterial filaments are more important for feeding at night when the reef waters contain much more plankton. However, based on the present study, little more can be said concerning the role of these filaments.

SUMMARY

1. A study was conducted of the chemical stimuli which elicit mouth opening and feeding behavior in the Hawaiian coral *Cyphastrea ocellina*. Investigated less intensively were *Fungia scutaria, Pocillopora damicornis, Tubastrea manni, Porites compressa, Pocillopora meandrina,* and *Leptastrea bottae*.

2. The feeding behavior of *Cyphastrea, Fungia,* and *Pocillopora* consisted of a wide mouth opening in response to live prey (plankters or *Artemia*), or to homogenates, or to seawater extracts of prey.

3. Eluates from chromatograms of alcoholic extracts of *Artemia* were tested on *Cyphastrea*. The major activity for mouth opening was largely restricted to a single spot, with somewhat less activity contained lower on the chromatogram. These spots consisted of the imino acid proline and the reduced tripeptide glutathione, respectively. These compounds were presumably released through wounds inflicted by the nematocysts of the corals in the body wall of the prey.

4. Using three methods of quantification, it was found that commercially prepared proline and reduced glutathione, as well as their respective analogs, pipecolic acid and S-methyl glutathione, caused strong mouth opening in *Cyphastrea*.

5. Commercial hydroxyproline also caused mouth opening, but this activity was thought to be due to contamination with small quantities of proline.

6. Plotting the number of *Cyphastrea* polyps out of 50 responding with time in different concentrations of proline, reduced glutathione, and S-methyl glutathione gave hyperbolic curves. A similar plot of the response to pipecolic acid was sigmoid in shape. The possible significance of this response to pipecolic acid is discussed.

7. The solitary coral *Fungia scutaria* and the branching colonial coral *Pocillopora damicornis* also responded to proline and reduced glutathione and occasionally to several other compounds.

8. *Cyphastrea* extruded more mesenterial filaments in glutathione whereas *Pocillopora* extruded more filaments in proline.

A NOTE ON THE GLUTATHIONE CONTROL OF FEEDING IN *CHLOROHYDRA VIRIDISSIMA*

In discussing the control of feeding in *Hydra littoralis*, Loomis (1955) stated that the glutathione feeding mechanism "is also present in the green hydra *Chlorohydra viridissima*, although in a less sensitive form." Many workers since, however, have been unable to observe an unequivocal feeding response to reduced glutathione in *Chlorohydra viridissima* (=*Hydra viridis*), although Rushforth (1965) found that reduced glutathione inhibited the light-induced contractions of *Hydra viridis*. Because I had observed that *Hydra pirardi, Hydra pseudoligactis, Hydra oligactis,* and *Hydra littoralis* respond to reduced glutathione, it seemed strange that *Chlorohydra viridissima* did not. My experiments revealed, however, that *C. viridissima*, like the other species of hydra tested, did respond to reduced glutathione, although in quite a different fashion than did *Hydra littoralis*.

For these experiments, two different groups of *Chlorohydra viridissima* were tested. One had been cultured in Hawaii for several years by Dr. L. V. Davis and the second in Florida by Dr. H. M. Lenhoff; both stocks were derived originally in Florida in 1961 from the same clone. The animals were raised in the laboratory in "M" solution (Muscatine 1961) and fed daily on freshly hatched nauplii of the brine shrimp *Artemia salina*. The hydra were not fed for 2 days before use.

Unlike the wide and sustained mouth opening of *Hydra littoralis* (Lenhoff 1961), *Chlorohydra viridissima* responded to *Artemia* nauplii and to *Artemia* homogenates and extracts by comparatively slight mouth opening—one approximately equal to the diameter of the oral cone. In addition, the mouth repeatedly opened and closed in a slow rhythmic fashion during the period of continuous chemical stimulation. Each successive mouth closure subsisted for longer and longer periods until finally the mouth did not reopen. Such a response made it extremely difficult to define the exact point at which *Chlorohydra viridissima* ceased to respond to chemical stimulation.

Using the methods for analyzing the extracts of *Artemia* as described in the main body of this report, I was able to show that the chemical contained in *Artemia* extracts which stimulated mouth opening of *Chlorohydra viridissima* was stable to boiling and soluble in 95 percent ethyl alcohol. The region of the chromatogram which contained the major activity for stimulating mouth opening had an R_F value of about 0.3. Among the compounds tested which had R_F values in this approximate area, commercially obtained reduced glutathione (Sigma Chemical Company, St. Louis, Mo.) gave a feeding response identical to that obtained with either live *Artemia* or *Artemia* extracts and homogenates. As mentioned previously, reduced glutathione has recently been shown to be abundant in *Artemia* nauplii.

Thus, it appears that *Chlorohydra viridissima*, like all other hydras tested to date, gives a feeding response to reduced glutathione. The response in *C. viridissima*, however, differs somewhat from that observed with *Hydra littoralis*. Such is the case with some of the other hydra species tested. *Hydra pirardi*, for example, may respond to reduced glutathione as long as 100 minutes at 22° C, its mouth closing and opening many times during that period (Lenhoff 1968). Another example, *Hydra pseudoligactis*, although responding to free glutathione, is observed occasionally to ingest inert material in the absence of added glutathione (Lenhoff 1968). Each species of hydra, therefore, may have its own peculiar feeding response to reduced glutathione.

LITERATURE CITED

Abe, N. 1938. Feeding behavior and the nematocyst of *Fungia* and 15 other species of corals. *Palao Tropical Biological Station Studies* 3: 469–521.

Benson, A. A., J. A. Bassham, M. Calvin, T. C. Goodale, V. A. Haas, and W. Stepka. 1950. The path of carbon in photosynthesis. V. Paper chromatography and radioautography of the products. *J. American Chemical Society* 72: 1710–1718.

Calvin, M. 1954. Mercaptans and disulfides: some physics, chemistry, and speculation. In *Glutathione,* S. Colowick et al., eds., pp. 3–30. New York: Academic Press.

Cohn, V. H., and J. Lyle. 1966. A fluorometric assay for glutathione. *Analytical Biochemistry* 14: 434–440.

Fulton, C. 1963. Proline control of the feeding reaction of *Cordylophora. J. General Physiology* 46: 823–837.

Isherwood, F. A. 1959. Chemistry and biochemistry of glutathione. In *Glutathione*, E. M. Cook, ed. *Biochemical Society Symposia* **17**: 3–16.
Kanazawa, A., Y. Kakimoto, T. Nakajima, and I. Sano. 1965. Identification of α-glutamylserine, α-glutamylalanine, α-glutamylvaline and S-methylglutathione of bovine brain. *Biochemica et biophysica acta* **111**: 90–95.
Lenhoff, H. M. 1961. Activation of the feeding reflex in *Hydra littoralis*. I. Role played by reduced glutathione, and quantitative assay of the feeding reflex. *J. General Physiology* **45**: 331–344.
———. 1967. Some ionic, chemical and endogenous factors affecting behavior of hydra. In *Chemistry of learning*, W. C. Corning and S. C. Ratner, eds. New York: Plenum Press.
———. 1968. Chemical perspectives on the feeding response, digestion, and nutrition of selected coelenterates. In *Chemical zoology*, vol. 2, M. Florkin and B. Scheer, eds., pp. 157–221. New York: Academic Press.
Lenhoff, H. M., and H. A. Schneiderman. 1959. The chemical control of feeding in the Portuguese Man-of-War, *Physalia physalis* L. and its bearing on the evolution of the Cnidaria. *Biological Bulletin, Woods Hole* **116**: 452–460.
Loomis, W. F. 1955. Glutathione control of the specific feeding reactions of hydra. *Annals, New York Academy of Sciences* **62**: 209–228.
Monod, J., J. Wyman, and J. -P. Changeux. 1965. A plausible model of allosteric transition. *J. Molecular Biology* **12**: 88–118.
Muscatine, L. 1961. Symbiosis in marine and freshwater coelenterates. In *The biology of hydra and of some other coelenterates: 1961*, H. M. Lenhoff and W. F. Loomis, eds., pp. 255–268. Coral Gables: University of Miami Press.
Rushforth, N. B. 1965. Inhibition of contraction responses of hydra. *American Zoologist* **5**: 505–513.
Wieland, T. 1954. Chemistry and properties of glutathione. In *Glutathione*, S. Colowick et al., eds., pp. 45–59. New York: Academic Press.

JOHN REES
University of Puerto Rico, Río Piedras

CHAPTER 11

Paths and Rates of Food Distribution in the Colonial Hydroid *Pennaria*

The paths and rates of food distribution in colonial hydroids are of particular interest for several reasons. For one, colonial hydroids have specialized polyps, some of which function to capture, ingest, and initiate the distribution of food to other polyps and to the nonfeeding portions of the colony. Hence, experiments on paths and rates of food movement may show how the activities of the polyps are coordinated. Secondly, modifications of these experiments might allow investigation of the distribution of nutriments during such developmental phenomena as stolon elongation, hydranth regression, and polyp formation.

To date most techniques used to investigate food distribution in colonial hydroids have utilized food labeled with dye. Strehler and Crowell (1961) followed the distribution in *Campanularia flexuosa* of ingested *Artemia* that had been stained with a fluorescent acridine dye. Among the coelenterates, radioactive tracers and autoradiography have been applied thus far primarily on *Hydra littoralis* (e.g., Lenhoff 1961), which normally are not colonial.

In this chapter we describe methods, and provide a few examples, of how radioactive food can be used to study the paths and rates of food distribution in the colonial hydroid *Pennaria tiarella*.

MATERIALS AND METHODS

Culture Conditions

Colonies of *Pennaria tiarella* (McCrady 1857) were collected from floating docks at the Hawaii Institute of Marine Biology. Small pieces of each colony were placed on slides with threads by a method similar to that described by Fulton (1960). The slides, in slide racks, were set in

---

An earlier version of this chapter appeared as "Paths and rates of food distribution in the colonial hydroid *Pennaria*" by John Rees, Lary V. Davis, and Howard M. Lenhoff, in *Comparative Biochemistry and Physiology* **34**: 309–316.

8"-diameter finger bowls containing artificial seawater ("Instant Ocean," Aquarium Systems, Inc., Wickliffe, Ohio), made up without the trace element additives. Specific gravity was maintained at 1.025. The colonies were fed to repletion once daily with *Artemia* nauplii and were constantly aerated. The finger bowls were cleaned of debris daily with a pipette, and the medium was changed every 3rd day. All cultures were maintained at ambient laboratory temperatures.

Experimental Procedures

The colonies used for experimentation were selected for similarity in size, shape, and number of hydranths. Each colony was isolated in a 4"-diameter finger bowl and starved for 12 hours prior to being used.

The labeled food used was kidney from a mouse that had been injected intraperitoneally with ^{14}C-protein hydrolysate dissolved in 0.1 ml physiological saline. Twenty-four hours after injection the mouse was killed and its kidney removed and frozen. The kidney was minced, soaked for several minutes in 10^{-3} M proline (see Pardy, this volume), and with fine forceps fed to the terminal hydranth.

At specific intervals after the colonies were given radioactive food, they were rinsed in seawater and were fixed on a 1"-diameter Millipore filter in the manner described for *Hydra littoralis* by Lenhoff (1959). After fixation, some colonies were used to make whole mount autoradiographs; others were cut into several pieces for counting, the pieces consisting of terminal hydranths, central stem segments, and lateral hydranths or lateral stem segments (see Figure 1). Each section was fixed to an aluminum planchet and counted in a Nuclear Chicago gas flow counter.

Autoradiography

The Millipore filter to which the colony was affixed was pasted onto a 1" × 3" glass slide with rubber cement. The slides were then covered with Saran Wrap, placed on Kodak blue sensitive medical X-ray film, wrapped in aluminum foil, and placed in the dark. Colonies registering 100 to 500 counts per minute on the gas flow counter gave good autoradiographs within a 3-day exposure period.

Protein Determination

Protein nitrogen was determined by the method of Lowry et al. (1951). The amount in each section of a colony was determined and is shown on Table 1. Note that the central stem contained more protein than did the lateral stem. Furthermore, the terminal hydranth, situated at the tip of the central stem, contained more protein than did the lateral hydranths.

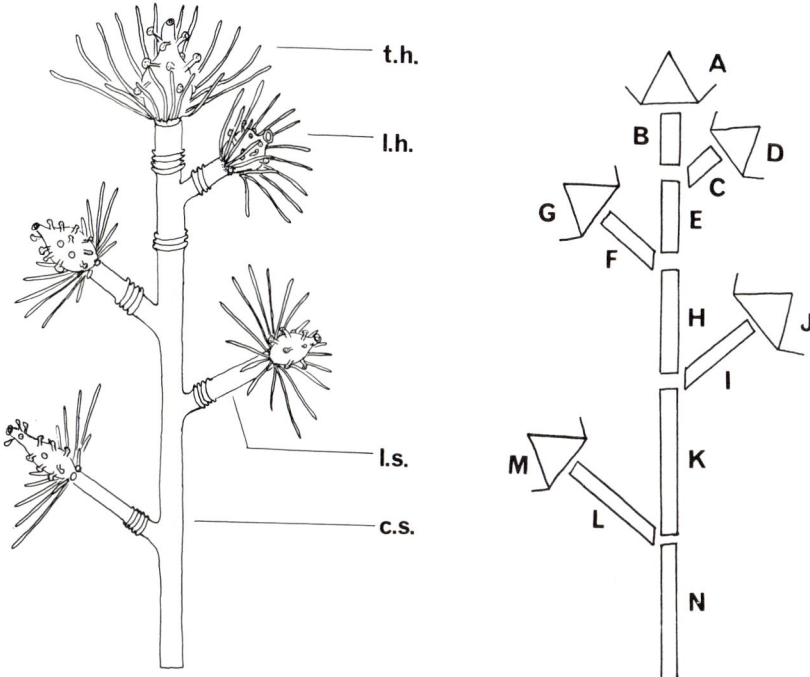

FIG. 1. Right, a *Pennaria tiarella* colony: *t.h.*, terminal hydranth; *l.h.*, lateral hydranth; *l.s.*, lateral stem; *c.s.*, central stem (hydrocaulus). Left, diagrammatic sketch of the same colony sectioned for counting.

RESULTS

Rate and Extent of Distribution of Label Fed to Terminal Hydranth

A terminal hydranth on a *Pennaria* colony was fed labeled tissue and was fixed 30 minutes later. Four pieces corresponding to [B, C, and D], [E, F, and G], [H, I, and J], and [K, L, and M] of Figure 1 were removed and counted. The results (Table 2, column 5) show that there was a

TABLE 1 Protein nitrogen in various sections of a *Pennaria* colony

Section*	Protein Content
Central stem	5.90 μg N/cm
Lateral stem	1.46 μg N/cm
Terminal hydranth	3.00 μg N/hydranth
Lateral hydranth	2.00 μg N/hydranth

NOTE: Protein determined by method of Lowry et al. (1951).
*Approximately 9 units of each section were used per measurement.

TABLE 2 Distribution of radioactivity in *Pennaria* colony 30 minutes after ingestion of food by terminal hydranth

Identity of Section (see Figure 1) and Median Distance from Terminal Hydranth (cm)	Counts per Minute per Section			Total 3 Sections
	Central Stem	Lateral Stem	Lateral Hydranth	
B 0.22	35.2	—	—	47.4
C 0.44	—	5.6	—	
D 0.56	—	—	6.6	
E 0.54	22.0	—	—	26.6
F 0.74	—	2.2	—	
G 0.86	—	—	2.4	
H 0.86	11.6	—	—	15.4
I 1.02	—	3.4	—	
J 1.29	—	—	0.4	
K 1.22	4.8	—	—	7.2
L 1.50	—	0.8	—	
M 1.68	—	—	1.6	
N 1.76	6.4			

gradient of radioactivity which decreased in proportion to the distance of the tissue from the terminal hydranth. The presence of activity in pieces [K, L, and M] shows that the ingested food was distributed rapidly throughout the colony.

Examination of the individual sections showed the central stem to be the most active (Table 2, column 2). This was to be expected because the central stem contained more protein (hence, probably more tissue) per millimeter than the other stem sections (Table 1), and was the main passageway for the food to reach the other sections. Gradients also were noticed in the distribution of radioactivity in the central stem sections, lateral stem sections, and lateral hydranths (Table 2, columns 2, 3, and 4).

Whole mount autoradiographs of a colony given labeled food at the terminal hydranth (Figure 2) were in accord with the results presented in Table 2. A definite gradient of radioactivity is apparent. The radioactivity is located in the coenosarc and possibly in the perisarc of stems and hydranths. The terminal hydranth is labeled heaviest, with progressively less radioactive labeling seen in the lateral hydranths as their distance from the terminal hydranth increases. That some of the tentacles on one of the first lateral hydranths on the right side are labeled indicates that the label is being passed directly through the tentacle cells because the tentacles of *Pennaria*, unlike those of hydras, are not hollow. The dense radioactivity in the hydrocoel of the stems is readily apparent; this activity, however, might also be due to the effect of the coenosarc that has dried and collapsed onto one side of the stem.

Transport of Label to the Basal Tip and to Growing Regions of the Colony

There was a slight increase in radioactivity in the basal region of the central stem (Table 2, section N), although it is not possible to tell whether this slight increase is significant. However, in another experiment not dealt with here in detail, autoradiographs were made of a colony that had a hydranth at each end of a stolon. These two hydranths were fed labeled mouse kidney exclusively for 4 days and then were fixed. The areas on the autoradiograph showing the most radioactivity were those representing two newly formed hydranths and the newly formed stolon. Gradients of radioactivity, however, were seen directly behind the original two "terminal" hydranths that had been fed the labeled food. The results of this experiment suggest that newly ingested food was used to form new hydranths and stolons distant from the hydranths that had ingested the food.

In another experiment, eight different colonies were given approximately equal amounts of radioactive food. At different intervals following administration of the label, one of the colonies was fixed and the radioactivity in the terminal hydranth and in the basal section (pieces

K, L, and M) was counted. Here (Table 3) the data are expressed as counts/min/μg protein nitrogen. The results show that, between 90 to 720 minutes, the amount of radioactivity in the section of the colony farthest from the terminal hydranth increased as the activity in the terminal hydranth decreased.

Effect of Food Ingested by Lateral Hydranths on the Distribution of Radioactive Food Ingested by the Terminal Hydranth

Immediately after the terminal hydranth was given a piece of labeled mouse tissue, the lateral hydranths were fed to repletion with unlabeled *Artemia* nauplii. At various intervals following feeding of the terminal hydranths, the colonies were fixed and prepared for counting. In general,

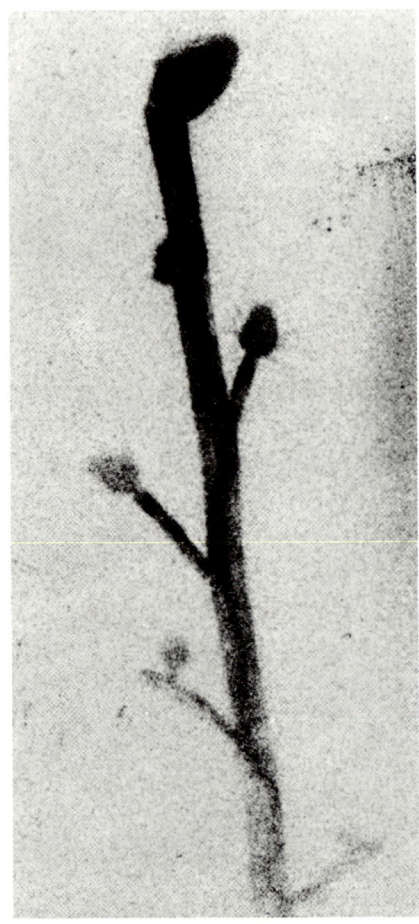

FIG. 2. Whole mount autoradiograph of *Pennaria tiarella*.

TABLE 3 Disappearance of label from terminal hydranth, and appearance of label in distal end of the colony

Time After Fixing Colony (minutes)	Counts/Minute/μg Protein Nitrogen	
	Terminal Hydranth	Distal Section [K, L, M]
10	63	0.0
20	73	0.0
30	21	0.0
90	51	1.5
180	45	2.0
360	14	8.0
540	20	12.0
720	4	13.0

the results indicate that more labeled food was retained in the terminal hydranth than was shown by the experiments in which the lateral hydranths were not given unlabeled food. For example, after 9 hours the terminal hydranth of the colony also given unlabeled food had 24 percent of the administered label, whereas the terminal hydranth from the colony not given unlabeled food had 6.4 percent of the label. Likewise, there seemed to be a delay in the appearance of the label in the central and lateral stems, presumably because of the effect of the unlabeled food coming from the lateral hydranths.

Table 4 shows the distribution of the radioactivity in the two experimental colonies fixed 9 hours after the terminal hydranth was fed labeled liver. In the colony in which only the terminal hydranth was fed, the radioactivity appeared more or less evenly distributed throughout the colony with somewhat more activity in sections away from the terminal

TABLE 4 Effect of food ingested by lateral hydranths on distribution of the food ingested by the terminal hydranths

Sections	Radioactivity (cpm/combined sections)	
	Lateral Hydranths Not Fed	Lateral Hydranths Fed Unlabeled Nauplii
B,C,D	70	107
E,F,G	–	32
H,I,J	67	34
K,L,M	109	20
N,O,P	93	–
Q,R,S	–	20

NOTE: The food given the terminal hydranth was radioactive mouse liver; the food given the lateral hydranths (immediately after the terminal hydranths ingested the liver) consisted of *Artemia* nauplii. The sections were fixed 9 hours after the hydranths were fed.

hydranth. On the other hand, in the experiment in which the lateral hydranths were given unlabeled food, by 9 hours the label was still heaviest in the terminal hydranth; the overall distribution gave a gradient similar to that observed in the experiment described in Table 2. In that experiment, the activity was counted 30 minutes after the terminal hydranth had been fed unlabeled food and the lateral hydranths had not been fed. Presumably the unlabeled food initially nourished the lateral hydranths and lateral stems, and in some way delayed the transport of labeled food to these areas from the terminal hydranth. Complications due to regression are mentioned in the discussion.

DISCUSSION

In this paper methods have been described showing how the paths and rates of food distribution can be investigated in the colonial hydroid *Pennaria tiarella*. Data are presented regarding gradients of labeled food, transport of food to growing regions, and transport of food particles under various conditions of feeding.

Although interpretation of the results seems straightforward, it should be kept in mind that the stolons are covered by a chitinous perisarc which may reduce the week emission of the ^{14}C label. Furthermore, the chitinous covering on the central stem is thicker than that on the lateral stem. The hydranths, on the other hand, are not covered by a perisarc.

The hydroplasm (fluid of the hydrocoel) of *Pennaria* stolons flows throughout the colony. The movement of the hydroplasm with the food particles it contains can readily be seen, with the aid of a phase contrast microscope, in a *Pennaria* colony that has just ingested a number of *Artemia* nauplii. The particles in the hydroplasm of the stolons were observed to move with great speed; the direction of movement was reversed frequently. Many authors have reported a similar movement of the hydroplasm in hydroids (e.g., Crowell 1957; Hale 1960). Fulton (1963), using time-lapse photography, showed waves of peristaltic contractions which began at the tips of the hydranths and moved toward the tips of the stolons in *Cordylophora lacustris*. After the food was ingested, the rate of contraction accelerated; after several hours, it declined to the resting state.

Strehler and Crowell (1961) showed that similar peristaltic movements occur in *Campanularia flexuosa*. In their studies, they followed the distribution of fluorescent dye from ingested stained *Artemia* nauplii. That these peristaltic contractions circulate nutrients efficiently was indicated by Crowell's experiments (1957) in which an unfed half of a colony of *Campanularia* lived for the same length of time as did the well-fed half of the same colony.

Most of the interpretations of the experiments described in this paper are presented briefly under "Results." However, rather than answering definitively the questions posed, these experiments bring up new ones. What, for example, will be the movement of food (and of labeled cells) during periods of intensive hydranth regression? The hydranths of *Pennaria* begin to regress when the colony does not feed frequently. In fact, some regression occurred in experiments (Table 4) in which the hydranths were not fed for 9 hours after the terminal hydranth had been fed. Regression did not occur after 9 hours, however, in the colony in which the lateral hydranths, as well as the terminal hydranth, were fed. Before regression takes place, food is probably incorporated in the coenosarc cells along a gradient (Table 2), but as cellular movement ensues during regression, the gradient is probably disrupted.

What factors control distribution of food? Does control depend upon which particular hydranths ingest the food? Why will feeding of lateral hydranths affect distribution of food from terminal hydranths? Is such control physical, i.e., does food from the lateral hydranths interfere with the mechanics of the flow of food from the terminal hydranth? Or is there an intracolonial regulatory factor in operation? If there is such a regulation, is it controlled in part by chemicals released from the ingested food? Tyrosine, for example, released from ingested food, aids in regulation of "neck"-contraction responses associated with feeding in hydra (Blanquet and Lenhoff 1968).

Of particular interest are the preliminary data showing relatively greater amounts of radioactivity in areas of growth, such as those areas where new stolons and new hydranths are being formed. Are the cells in these areas selectively taking more food for cell division out of the hydroplasm? Are labeled cells from other parts of the colony migrating to areas of new growth?

These methods might be applied to an investigation of such aspects of the biology of *Pennaria* as chemistry of metabolic pools, cellular turnover, and aging.

SUMMARY

1. Methods are described for measuring the rate and paths of distribution of radioactive food in the colonial hydroid *Pennaria tiarella*.

2. Food fed to the terminal hydranth was found to be distributed throughout the entire *Pennaria* colony within 30 minutes after it was ingested. At that time the radioactivity appeared as a gradient, decreasing in proportion to the distance of the tissue from the terminal hydranth. The central stem was the most radioactive part of the colony.

3. In growing colonies, radioactive food fed to the terminal hydranths seemed to be preferentially utilized by the growing regions.

4. When unlabeled food was ingested by the lateral hydranths at

about the same time the terminal hydranth ingested radioactive food, the distribution of radioactive food into the lateral stems and hydranths was delayed.

5. Suggestions are made for the application of these methods to further studies of food distribution and of development in colonial hydroids.

LITERATURE CITED

Blanquet, R. S., and H. M. Lenhoff. 1968. Tyrosine enteroreceptor of hydra: Its function in eliciting a behavior modification. *Science* **159**: 633–634.

Crowell, S. 1957. Differential responses of growth zones to nutritional level, age, and temperature in the colonial hydroid *Campanularia*. *J. Experimental Zoology* **134**: 63–90.

Fulton, C. 1960. Culture of a colonial hydroid under controlled conditions. *Science* **132**: 473–474.

———. 1963. Rhythmic movements in *Cordylophora*. *J. Cellular and Comparative Physiology* **61**: 39–51.

Hale, L. J. 1960. Contractility and hydroplasmic movements in the hydroid *Clytia johnstoni*. *Quaterly J. Microscopical Science* **101**: 339–350.

Lenhoff, H. M. 1959. Migration of ^{14}C-labeled cnidoblasts. *Experimental Cell Research* **17**: 570–571.

———. 1961. Digestion of protein in *Hydra* as studied using radioautography and fractionation by differential solubilities. *Experimental Cell Research* **23**: 335–353.

Lowry, D. H., N. J. Rosebrough, A. L. Farr, and R. J. Randall. 1951. Protein measurement with the Folin phenol reagent. *J. Biological Chemistry* **193**: 265–275.

Strehler, B. L., and S. Crowell. 1961. Studies on comparative physiology of aging. *Gerontologia* **5**: 1–8.

L. H. DiSALVO
University of North Carolina, Chapel Hill

CHAPTER 12

Ingestion and Assimilation of Bacteria by Two Scleractinian Coral Species

Bacteria in aquatic ecosystems serve primarily as agents of nutrient regeneration; however, in certain habitats where their growth is highly favored, they may serve as food for several types of suspension-feeding invertebrates (ZoBell and Feltham 1938; see Jørgensen 1966). There are apparently no reports of bacterial feeding by corals, although corals live in close proximity to large numbers of bacteria contained within the regenerative spaces of coral reefs (DiSalvo 1969b). Abe (1938) reported the ability of several species of corals to feed on lamellibranch eggs. More recently Roushdy and Hansen (1961) reported that the octocoral *Alcyonium digitatum* filters ^{14}C-labeled *Skeletonema costatum* (a phytoplankter) from seawater, and that ^{14}C-labeled material is recoverable from a tissue homogenate of the coral.

I attempted to determine the degree of bacterial fouling of living and dead coral surfaces, and to observe some general aspects of particle feeding by these corals. In addition, I presented two coral species with ^{35}S-labeled bacteria to determine if the ingested bacteria were assimilated by the coral polyps.

MATERIALS AND METHODS

The corals studied included *Fungia scutaria* and *Pocillopora damicornis*. They were collected on a mid-Kaneohe Bay patch reef and maintained in running seawater for up to 3 days before experimentation. Observations on particle movement and feeding by the corals were made both with the unaided eye and with a binocular stereoscopic microscope at about 20X.

Coral surfaces were assayed for bacteria by gently pressing the well-drained coral surface onto nutrient agar plates. The plates were examined for bacterial growth along the coral imprint lines after 9-, 16-, and 24-hour incubations at room temperature. To sterilize their surfaces, some *Fungia* specimens were soaked for periods up to 36 hours in an antibiotic solution including 100 units/ml penicillin G, 20 µg/ml polymyxin B, and 10 µg/ml tetracycline in 0.45 µ (pore size) Millipore-filtered seawater (MFSW).

Ten presumably different species of bacteria were isolated from coral surface imprint plates for subsequent feeding studies. These species were selected on the basis of macroscopically visible differences based on colony pigmentation, morphology, and growth characteristics. Bacteria were isolated and subcultured using ZoBell's medium no. 2216 (ZoBell 1946) containing 1.5 percent agar, 5 percent peptone, 1 percent dextrose, 0.01 percent $FePO_4$ and 75 percent seawater. Agar plates were poured in Petri dishes, or liquid cultures were prepared in flasks by omitting the agar.

One bacterial isolate was labeled with ^{35}S. These bacteria, like many marine ones, were gram negative, asporogenous, and rod-shaped ($2 \times 4\,\mu$). They were inoculated into 10 ml of the liquid medium which contained a neutralized acid hydrolysate of 50 mg of ^{35}S-labeled yeast cells (500 mc/mg, Schwartz Bio Research, Orangeburg, N.Y.), and were cultured for 5 days. The suspension was centrifuged at about $9,000 \times g$ for 20 minutes and the resultant pellet was resuspended in 20 ml MFSW. These processes were repeated twice, but after the last centrifugation, the bacteria were resuspended in 2 ml MFSW. A sample of bacteria was filtered, dried, and weighed, and the radioactivity of the sample was determined.

In order to induce corals to ingest the labeled bacteria, a suspension of homogenized plankton was mixed with the bacteria. The extract behaved as a physical carrier for the bacterial cells and presumably contained the feeding activator(s) (see Mariscal, this vol.) necessary to induce ingestion of the material entrapped in the coral mucus. To prepare the "plankton particles," about 5 grams of mixed zooplankters collected under a night light were ground for 5 minutes in 6 ml of seawater with a mortar and pestle. The mixture was centrifuged for 1 minute at about $50 \times g$; a heavy residue settled and was discarded. The suspension used for feeding the corals consisted of one part ^{35}S-labeled bacterial suspension to three parts plankton particle suspension (v/v). This feeding mixture contained 7.4×10^5 cpm/ml.

Corals were fed pure bacterial isolates or the labeled suspension in the following manner. The corals were maintained in 150-ml laboratory dishes, and, beginning late in the evening when the polyps were fully extended, were twice offered a particulate suspension. A pipette was used to release small clouds (50–100 µl) of this suspension gently over each organism or colony. Each feeding was administered over a 1-hour period, during which time the observations of feeding behavior were made. The corals fed labeled material were exposed to the suspension for an additional hour, after which time they were placed in filtered seawater for 24 hours. They were then removed, shaken free of excess water, and pressed onto nutrient agar plates. Finally, the corals were rinsed with tap water and were kept frozen until they were fractionated.

The bacteria on the imprint plates were allowed to grow for 12 hours; the entire surface of the agar was covered with a filter paper disc and allowed to incubate 6 hours more. The filter paper was then removed, dried in a vacuum dessicator, and autoradiographed to ascertain if any of the labeled bacteria fed to the corals remained on their surfaces.

Fractionations of the labeled coral tissue and of the radioactive bacteria were done using a modified Schmidt-Thanhauser technique (Lenhoff 1961). Before the fractionation, each coral specimen was thoroughly ground with a mortar and pestle and mixed with enough deionized water to produce a watery suspension.

All samples were plated on aluminum planchets lined with a disc of lens paper, and were evaporated to dryness for counting. Determinations of radioactivity were made using a Nuclear-Chicago thin window gas flow detector. Counts recorded are corrected for the approximate efficiency of the counting apparatus and background, although not for self-absorption and decay of the ^{35}S.

For chromatographic analysis, a sample of labeled bacteria was extracted with 80 percent ethyl alcohol for 30 minutes at 45° C, and the remainder was hydrolyzed in 6 N HCl in a sealed glass tube at 100° C for 24 hours. Both these fractions were chromatographed using descending two-dimensional paper chromatography with a phenol-water first phase and butanol-propionic acid-water second phase (Bassham and Calvin 1957). Autoradiographs of the chromatograms were made with Kodak X-ray film, allowing exposure times of 4 days. Spots on the autoradiograph were tentatively identified by comparison with maps published by Roberts et al. (1957), and Bassham and Calvin (1957).

RESULTS

Observations of Intact Corals

The ciliary currents and particle feeding in *Fungia* specimens were similar to those described by Abe (1938). In addition, I noted that when the plankton particles reached the small tentacles on the outer margins of specimens of *Fungia scutaria*, the coral no longer rejected those particles, and the direction of particle movement was reversed. Relatively turbid clouds of bacteria added to the corals were almost completely removed from the water in 1 to 3 minutes by mucus entrapment. Two minutes later, through subsequent ciliary movement of the settled particles, mucus strands were formed and were swallowed by the coral. Within ½ to 1 hour after the material was swallowed, a bolus was regurgitated. This series of events occurred with whole plankton, plankton particles, and pure bacterial suspensions. Small volumes of plankton or plankton particles (formed from a 50-100-μl suspension) were retained by the corals for periods much greater than 1 hour.

The mouth-opening and movement phases of the feeding response of

Fungia were inhibited by the antibiotic solution (described in "Materials and Methods"), whereas the secretion of mucus and the movement by ciliary action of trapped particles to the mouth did not appear to be affected by the antibiotics. Specimens of *Porites compressa* showed similar ciliary cleaning currents as described for several *Porites* species by Yonge (1930). I also noticed that occasionally this coral could trap particles in mucus and ingest those particles with the aid of ciliary movements. Although specimens of *Pocillopora damicornis* were observed to feed on plankton particles, they did not appear to trap the particles in mucus first.

Bacteria on the Surface of *Fungia*

Some coral surfaces showed extensive bacterial coverage while others were almost bacteria-free. I could discern no predictable pattern, although my impression was that *Fungia* periodically cleans its surface. This view was supported by field observations, during which I was often able to see large specimens ridding themselves of sedimentary accumulations by mucus secretion and removal. That some *Fungia* specimens had surfaces practically free of bacteria suggests that the mucus cleaning process also effectively removes bacteria. Nonliving coral skeletal surfaces were always heavily coated with bacteria.

Suspensions of the 10 bacterial isolates taken from coral surfaces and offered separately to specimens of *Fungia scutaria* were not readily retained as food. Only one bacterial suspension was retained by *Fungia*, although the bolus formed from this suspension, regardless of its size, was regurgitated 45 minutes after it was swallowed.

Fate of Radioactive Bacteria Ingested by Corals

Two specimens of *Fungia scutaria* about 2 cm in diameter and one 5-cm branch of *Pocillopora damicornis* were fed ^{35}S-labeled bacteria in the plankton homogenate. One day later the corals were fractionated with trichloroacetic acid (TCA) and ethyl alcohol, and the distribution of radioactivity within each fraction was compared to that of the labeled bacteria. Table 1 shows that most of the label in the bacteria was in the

TABLE 1 Distribution by percent of ^{35}S in major chemical fractions of the marine bacterial species which were used for feeding the corals

Fraction	Radioactivity (cpm)	Percentage of ^{35}S Recovered
TCA-Soluble	1.65×10^6	19.4
Alcohol-Soluble	4.89×10^4	0.6
TCA- and Alcohol-Insoluble	6.71×10^6	80.0

NOTE: Of the total radioactivity in the bacteria fractionated, 103 percent was recovered during the fractionation procedures.

protein fraction (TCA- and alcohol-insoluble). Autoradiographs of chromatograms of acid hydrolysates of the bacterial protein showed most of the radioactivity to be in methionine and cysteine and their derivatives.

In contrast, analysis of coral tissue 24 hours after the ingestion of labeled bacteria showed the amino acid fraction (TCA-soluble) to be the most active. Furthermore, the *Pocillopora* specimen had a relatively radioactive peptide fraction (TCA-insoluble, alcohol-soluble). Hence, it appears that not only were the labeled bacteria ingested by the corals, but the ingested bacteria were degraded and most likely assimilated. No radioactive bacteria were ever found remaining on the coral surfaces by the plating and autoradiographic technic.

DISCUSSION

Bacteria are convenient to use for particle-uptake studies in corals. They may have special ecological significance because they usually coat most particulate detritus in the sea as well as corals and most other reef surfaces (DiSalvo 1969a). That the degree of bacterial coating of corals varies considerably suggests a dynamic relationship in which the corals periodically shed or harvest their bacterial coating. There has been no direct experimental evidence that corals feed on bacteria. While swimming on the reef during the day, I have occasionally observed *Fungia* regurgitating material which appeared to be detrital in character, rather than of planktonic origin. Bacteria should not be considered as a purely energy-dissipating part of the ecosystem, because they are often 50 percent efficient in converting nutrients into protein (Brock 1966). For completion of mineralization cycles, the microbiological agents themselves must be broken down and recycled.

I have demonstrated in these experiments that, when corals are stimulated to ingest radioactive bacteria (mixed with plankton homogenate), the bacteria are degraded within the coral. The fractionation data show that much of the bacterial protein is rendered soluble. It is not certain whether the TCA-insoluble material observed in the coral is newly synthesized coral protein or unhydrolyzed bacterial protein. Whether from coral protein or bacterial protein, however, the new amino acids obtained from the hydrolyzed bacteria are probably mixed with the coral amino acids and are available for protein synthesis. Some of my more recent data have shown that the radioactive material in the corals rapidly changes from an insoluble to a soluble form. This soluble material then disappears and TCA-insoluble radioactive components arise. The amount of this insoluble material does not appear to change for more than 72 hours. I have also demonstrated that the dominant Kaneohe Bay coral species *Porites compressa, Montipora verrucosa, Pavona varians,* and *Pocillopora meandrina* have the ability to feed upon and digest the labeled bacteria (DiSalvo 1969a).

TABLE 2 Distribution of radioactivity in major chemical fractions of the tissues from corals fed labeled bacteria

Fraction	Fungia scutaria No. 1		Fungia scutaria No. 2		Pocillopora damicornis	
	Radioactivity (cpm)	% of Total Recovered	Radioactivity (cpm)	% of Total Recovered	Radioactivity (cpm)	% of Total Recovered
TCA-soluble	1.25×10^4	66.8	1.09×10^5	76.1	3.26×10^5	57.0
Alcohol-soluble	3.2×10^3	1.9	2.22×10^3	1.6	9.41×10^4	16.5
TCA- and Alcohol-insoluble	4.56×10^4	26.8	3.09×10^4	21.6	1.50×10^5	26.4

NOTE: The fractionations were carried out 24 hours after the corals were fed ^{35}S-labeled bacteria. Radioactivity recovered in the coral fractions was 94.2, 71.4, and 78 percent, respectively.

The differences in percent distribution of the label in *Fungia* and *Pocillopora* (Table 2) may reflect differences in their rates of assimilation, or may possibly indicate some basic metabolic difference between the two species. The overall importance of bacteria as food for corals has not been and cannot be evaluated from the present preliminary study, but the methods described herein may be of value in determining if such a nutritional role exists.

More important than the role of bacteria in the nutrition of corals, however, is the probability that the digestive activities of the corals on bacteria are an effective defense mechanism against microbial invasion. Phillips (1963) suggested the possibility that foreign particles might be phagocytized within the cnidarian coelentron, but presented no supportive evidence.

Several aspects of the feeding mechanisms of corals have been noted during these experiments. Of special interest is the reversal of particle direction that takes place as the particles reach the outer fringe of *Fungia* tentacles. Such a reversal indicates that there is a chemoreceptor distal to the mouth which may be part of some control system functioning to convert the particle-rejection response to particle-intake response. A related situation is cited by Jørgensen (1966:6) in his discussion of particle feeding by scyphozoan medusae. The studies he reviewed indicate that there are chemoreceptors far from the mouth which are capable of directing food particles to the mouth via ciliary tracts.

SUMMARY

Some observations are made on selected coral species concerning the means by which they ingest particles, and the nature of the bacteria living on their surfaces. Although sometimes the surfaces of living corals were relatively free of bacteria, they usually had a rich surface flora. None of the corals would retain several species of washed bacterial cells. Instead they trapped these cells with mucus and moved them by ciliary action. When they did swallow these cells, they regurgitated them within an hour. A particle extract of plankton fed to *Fungia scutaria* was handled much in the same way as was a bacterial suspension, the major difference being that the coral would ingest and retain small amounts of the plankton particles in the coelenteron. Bacteria labeled with ^{35}S were mixed with the particle plankton extract and were fed in small amounts to *Fungia scutaria* and *Pocillopora damicornis*. When the coral tissues were fractionated 24 hours later, they gave evidence that they had hydrolyzed and assimilated the bacteria.

LITERATURE CITED

Abe, N. 1938. Feeding behavior and the nematocysts of *Fungia* and 15 other species of corals. Contribution 22. *Palao Tropical Biological Station Studies* **3**: 469–521.
Bassham, J. A., and M. Calvin. 1957. *The path of carbon in photosynthesis.* Englewood Cliffs, N.J.: Prentice-Hall.
Brock, T. D. 1966. *Principles of microbial ecology.* Englewood Cliffs, N.J.: Prentice-Hall.
DiSalvo, L. H. 1969*a*. Regeneration functions and microbial ecology of coral reefs. Ph.D. dissertation, University of North Carolina.
―――. 1969. On the existence of a coral reef regenerative sediment. *Pacific Science* **23**: 129.
Jørgensen, C. B. 1966. *Biology of suspension feeding.* New York: Pergamon Press.
Lenhoff, H. M. 1961. Digestion of protein in *Hydra* as studied using radioautography and fractionation by differential solubilities. *Experimental Cell Research* **23**: 335–353.
Phillips, J. H. 1963. Immune mechanisms in the phylum Coelenterata. In *The lower Metazoa, comparative biology and phylogeny*, E. C. Doughterty, ed. Berkeley and Los Angeles: University of California Press.
Roberts, R. B., P. H. Abelson, D. B. Cowie, E. T. Bolton, and R. J. Britten. 1957. Studies on the biosynthesis of *Excherichia coli,* Publication 607. Carnegie Institution of Washington, D. C.
Roushdy, H. M., and V. K. Hansen. 1961. The filtration of phytoplankton by the octocoral *Alcyonium digitatum. Nature* **190**: 649–650.
Yonge, C. M. 1930. Studies on the physiology of corals. I. Feeding mechanisms and food. *Scientific Reports of the Great Barrier Reef Expedition,* 1928–1929:**1** (2).
ZoBell, E. E. 1946. *Marine microbiology.* Waltham, Mass.: Chronica Botanica Publishing Co.
ZoBell, E. E., and C. Feltham. 1938. Bacteria as food for certain marine invertebrates. *J. Marine Research* **1**: 312–327.

GORDON R. MURDOCK
Duke University, Durham, North Carolina

CHAPTER 13

The Formation and Assimilation of Alcohol-Soluble Proteins during Intracellular Digestion by *Hydra littoralis* and *Aiptasia* sp.

Digestion in coelenterates is thought to occur in two steps. First, food is broken into small particles by enzymes in the lumen of the coelenteron; second, the particles, after being phagocytized by cells lining the coelenteron, are completely digested within food vacuoles. These two steps, shown by both cytological and chemical methods (Gauthier 1963; Lenhoff 1968), are termed respectively extracellular and intracellular digestion.

Despite the well-documented literature on intracellular digestion in coelenterates, we have little information on the chemical changes actually taking place within the food vacuole. In the present paper I describe observations on *Hydra littoralis** and on *Aiptasia* sp. in which the labeled protein of ingested food is degraded intracellularly into a form which is precipitable in 5 percent trichloroacetic acid (TCA), yet is soluble in warm 80 percent ethanol. As this material—which is probably a large peptide or fragment of a partially degraded protein—disappears, new coelenterate proteins appear.

MATERIALS AND METHODS

Culture of Animals

Hydra littoralis was grown in mass culture by the methods of Loomis and Lenhoff (1956) in a solution of 10^{-3} M CaCl$_2$ and 10^{-4} M NaHCO$_3$, pH 7.4. It kept without food 2 days prior to the experiment.

Sea anemones of *Aiptasia* sp., were collected in Kaneohe Bay. To avoid complications resulting from the presence of the endosymbiotic zooxanthellae, the animals were placed in the dark at 37° C for 1 or 2 weeks and were fed daily on nauplii of the brine shrimp *Artemia salina*.

An earlier version of this chapter appeared as "Alcohol soluble proteins: their formation and assimilation during intracellular digestion in *Hydra littoralis* and *Aiptasia* species" by Gordon R. Murdock and Howard M. Lenhoff, in *Comparative Biochemistry and Physiology* **26** (3): 963–970.

*The experiments on *Hydra littoralis* were carried out by H. Lenhoff.

Under these conditions most of the anemones turned pale brown to white, having lost most of their algae. After this treatment, the anemones were maintained in the laboratory in seawater on a daily diet of *Artemia* nauplii for about a week prior to the experiment.

Types of Radioactive Food

The radioactive food fed to the *Hydra* was ^{35}S-labeled mouse lung; the mouse had been made radioactive by injecting it intraperitoneally with a neutralized hydrolysate of *Escherichia coli* grown on ^{35}S-sulfate in physiological saline.

The radioactive food fed to the anemones was ^{14}C-labeled mouse kidney. The mouse was injected intraperitoneally with 0.5 mc of a ^{14}C-labeled protein hydrolysate dissolved in 0.1 ml physiological saline. After 24 hours following injection, the mouse was killed and dissected. Its organs were frozen until used.

Fractionation of Tissues

The fractionations of the tissues by their differential solubilities in TCA and ethanol were carried out as described in Roberts et al. (1955) and in Lenhoff (1961). In the ^{35}S experiments, the radioactivity was distributed as follows: the cold TCA-soluble (TCA-S) fraction contained the amino acids; the cold TCA-insoluble, alcohol-soluble (Alc-S) fraction contained large peptides or small proteins; the cold TCA-insoluble, alcohol-insoluble (Alc-I) fraction contained protein. In the ^{14}C experiments, the Alc-I fraction was further fractionated into an alcohol-insoluble, hot TCA-soluble (HTCA-S) portion, which contained nucleic acid, and an alcohol-insoluble, hot TCA-insoluble (HTCA-I) portion which contained protein.

Analyses of Radioactive Food

Fractionation of ^{35}S-labeled mouse lung showed about 23 percent of the total radioactivity in the TCA-S and 77 percent in the protein (TCA-precipitable). Of the protein fraction, two-thirds of the label was in the Alc-I fraction, while one-third was in the Alc-S. This large proportion of label in the Alc-S fraction was shown to be of a protein (or peptide) nature and not lipid, because it did not migrate on a paper chromatogram suspended in a butanol-formic acid solvent. Analysis of a hydrolysis of this fraction showed the radioactivity to be distributed between cysteine-cystine and methionine. It should be noted that tissues high in Alc-S proteins are not always obtained in labeled mice (Lenhoff 1961).

The distribution of radioactivity among the various fractions of the ^{14}C-labeled mouse kidney analyzed is expressed as the percentage of the total radioactivity present in each piece analyzed. The TCA-S fraction contained from 1 to 7 percent of the total radioactivity; the Alc-S, from 7

Formation and Assimilation of Proteins during Intracellular Digestion

to 18 percent; the HTCA-S about 1 percent; and the HTCA-I, from 74 to 91 percent. The variation could be due to (*a*) irregular labeling of the different anatomical parts of the kidney that happened to be used for fractionation; (*b*) some of the more soluble components leaking out of the tissue before it was fractionated; or, simply, (*c*) slight contamination during the fractionation procedure.

Removal from Coelenteron of Radioactive Food Not Taken Up

Radioactive food that was not taken up by *Hydra*'s gastrodermal cells was removed from its gastrovascular cavity by the bisection-washing procedure described by Lenhoff (1961). Such food was removed from the anemones simply by flushing out the gut with seawater at the appropriate time; this procedure was followed with all anemones killed within 24 hours after they had been fed the radioactive tissue.

RESULTS AND DISCUSSION

Experiments with *Hydra*

Usually 10 to 12 individuals of *Hydra littoralis* were used per experiment. Each *Hydra* was fed small bits of ^{35}S-labeled mouse lung dipped in reduced glutathione (Lenhoff 1961). At hourly intervals following ingestion, the radioactive food remaining in the gut was removed (see "Materials and Methods") and the animals' tissues were fractionated. The guts of *Hydra* that had regurgitated their wastes (usually within 6 hours after they were fed) were not washed. The *Hydra* tissue was fractionated at varying times following ingestion of the radioactive food. The results are shown in Figure 1.

At 1-2 hours after ingestion, the food phagocytized by the gastrodermal cells did not differ significantly in the distribution of label among the major chemical fractions from that of the labeled food before it was ingested. Hence, it was concluded that the bulk of the food protein was not hydrolyzed into small molecules in the lumen of the gut, but was phagocytized by the gastrodermal cells.

Within 7 hours after the food was phagocytized, the Alc-S fraction doubled, while there was a concomitant decrease in the Alc-I fraction; no measurable change occurred in the TCA-S fraction. These changes probably represent the intracellular degradation of the food protein into smaller proteins and/or large peptides that are soluble in alcohol.

That this degradation of Alc-I into Alc-S occurred intracellularly was also shown by analyzing the food remaining in the gut lumen 3 hours after ingestion. Such food had the same properties of Alc-I and Alc-S as the original food. On the other hand, by 3 hours following ingestion, the proportion of label in the Alc-S fraction of the food within the gastrodermal cells had nearly doubled.

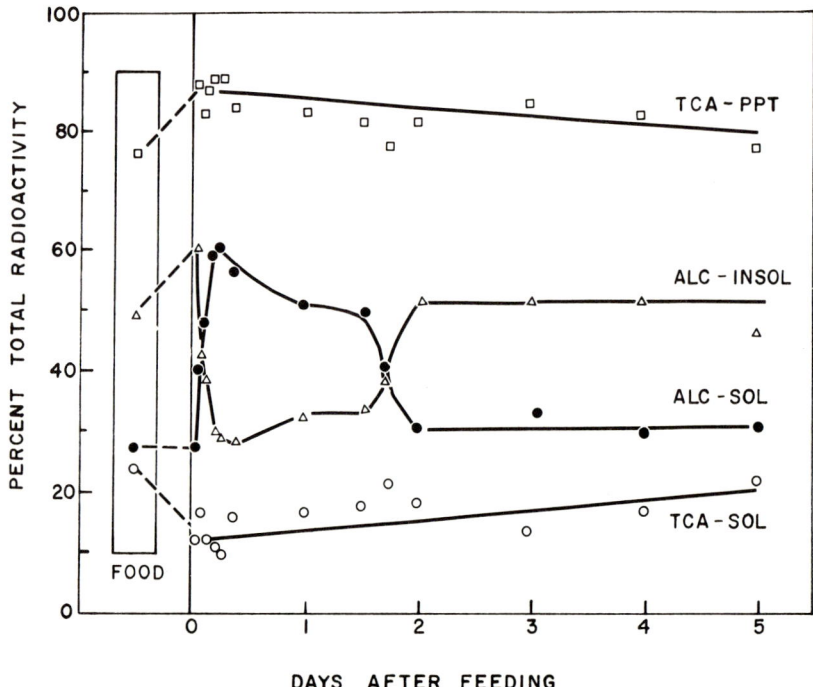

FIG. 1. Fate of ^{35}S-labeled tissue ingested by *Hydra littoralis.* In the bar on the left of the figure is shown the percentage distribution of TCA-precipitable label in the food (□); TCA-precipitable and insoluble in alcohol (△); TCA-precipitable and soluble in alcohol (●); and TCA-soluble (o). The curves on the right of the figure show the proportion of label in each of these fractions after the food has been taken up by the *Hydra's* gastrodermal cells.

Once the proportion of the label in the Alc-S fraction increased, it remained at that high level for about 1½ days. During the next 12 hours, however, the Alc-S fraction fell to about 30 percent of the total radioactivity and remained constant at that level for at least 5 days. Simultaneous with the changes in the amount of Alc-S fraction, the Alc-I increased until it constituted almost 60 percent of the radioactivity.

These changes in the Alc-S and Alc-I fractions indicate that during the 1½ days following ingestion the Alc-S protein was slowly degraded to amino acids to be used for synthesis of *Hydra* protein. A more rapid hydrolysis of the Alc-S fraction presumably occurred from 1½ to 2 days, and is reflected in the rapid increase of label in hydra protein (Alc-I) that occurred at that time. An increase in the amino acids resulting from the rapid hydrolysis of the Alc-S fraction might be indicated by the transitory increase of label in the TCA-S fraction and decrease of label in the TCA precipitate which was observed at 1-3/4 days.

Experiments with *Aiptasia* sp.

In these experiments the animals were fed the label in one of two ways: (*a*) an *Aiptasia* was fed a small bit of labeled mouse kidney followed immediately by a few *Artemia* nauplii, or (*b*) a homogenate made from a piece of mouse kidney and *Artemia* extract (fluid from *Artemia* homogenized in seawater) was pipetted directly into the coelenteron through the mouth.

Fractionation of *Aiptasia* sp.

At various times after single *Aiptasia* were fed ^{14}C-labeled mouse kidney along with unlabeled *Artemia* (whole or extract), the *Aiptasia* were fractioned. In addition to measuring the radioactivity in the cold-TCA-soluble and alcohol-soluble fractions as I did in the hydra experiments, I further fractionated the alcohol-insoluble material into hot TCA-soluble (HTCA-S) and hot TCA-insoluble (HTCA-I) fractions. Because ^{14}C label was used, the HTCA-S was presumed to be nucleic acid and the HTCA-I, protein. The results (Figure 2) show that, once the labeled tissue was taken up by the anemone, the radioactivity in the Alc-S fraction more than doubled within 2–3 hours, until it made up from 30 to 40 percent of the radioactive material within the anemone. Soon after this high Alc-S fraction was formed, however, it began to disappear at what seemed to be a first order rate until, by 100 hours after ingestion, it constituted about 5 percent of the label in the anemone tissue.

These findings are interpreted to indicate that, once inside the anemone cells, the labeled food protein is first partially degraded into large peptides (Alc-S), which are then further degraded into amino acid for anemone protein synthesis.

A comparison of the two curves drawn on Figure 2 shows that label was taken up into the anemone protein fraction (HTCA-I) as it was lost from the Alc-S. In Figure 2*b* it can be seen that within 1 hour after the labeled tissue was taken up by the anemone, the proportion of label in the protein fraction dropped to about 30 percent of the total. Following this rapid decrease in protein fraction, there followed an immediate, steady increase of label in this fraction until by 100 hours after ingestion it comprised about 90 percent of the label in the anemone.

I regard the results in Figure 2*b* to be in accord with the interpretation of Figure 2*a*. It should be noticed, nonetheless, that (*a*) not all of the radioactivity lost from the HTCA-I protein fraction could be accounted for by the increase in radioactivity in the Alc-S fraction; (*b*) the anemone HTCA-I protein began to increase before the maximum amount of Alc-S material had formed.

These observations are explicable if we take into account the changes that occurred within the cold TCA-S fraction (Figure 3*b*). The results showed that within an hour after the animals ingested labeled food, the

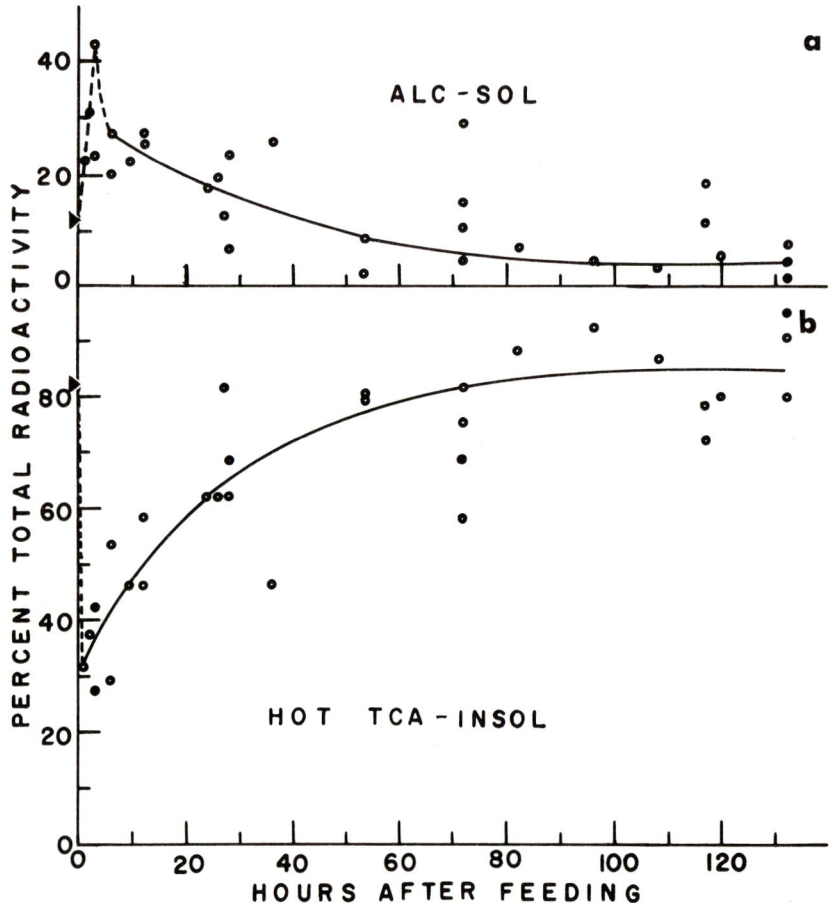

FIG. 2. Fate of tissue labeled with ^{14}C amino acids after being ingested by *Aiptasia* sp. Curve a shows a labeled material insoluble in TCA but soluble in alcohol, expressed as a percentage of the total radioactivity in *Aiptasia* tissues. Curve b shows labeled material insoluble in cold TCA, in alcohol, and in hot TCA, expressed as in curve a. The arrows (▶) indicate mean values for the percentage of label in each fraction of the food prior to ingestion.

label in the cold TCA-S material increased until it constituted nearly 40 percent of the total label in the animal.

Thus, these results indicate that some of the ingested protein was rapidly hydrolyzed to amino acids. These amino acids presumably contributed directly to synthesis of the new anemone protein which began forming within 2 hours following ingestion of the food (Figure 2b).

Examination of the changes in the nucleic acid fraction (HTCA-S) (Figure 3a) shows no significant increase or decrease. The two high points

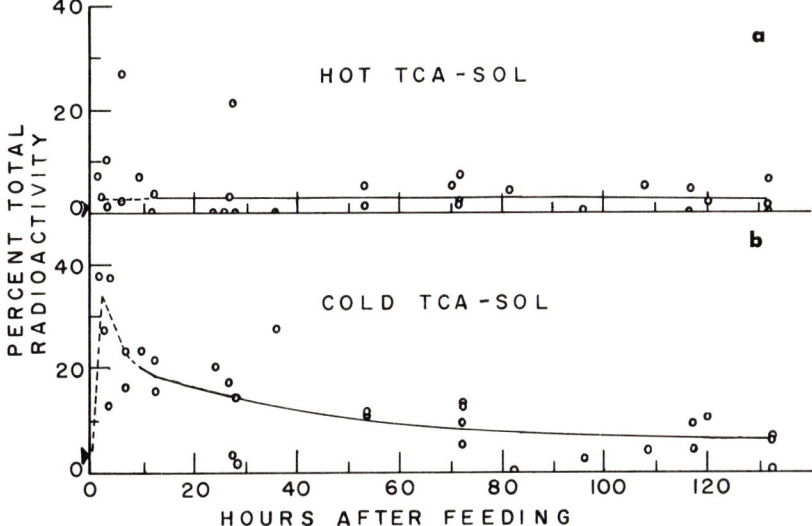

FIG. 3. Fate of tissue labeled with ^{14}C amino acids after being ingested by *Aiptasia* sp. Curve *a* shows a labeled material insoluble in both cold TCA and alcohol, but soluble in hot TCA, expressed as a percentage of the total radioactivity in *Aiptasia* tissues. Curve *b* shows labeled material soluble in TCA, expressed as in curve *a*. The arrows (▶) indicate mean values for the percentage of label in each fraction of the food prior to ingestion.

observed at 6 and 28 hours are probably the results of contamination with label from other fractions. Significantly, the results do indicate that, under these conditions, if any new nucleic acids are formed during the 100 hours following ingestion, amino acids from the ingested labeled protein did not contribute carbon. Presumably, if anemone nucleic acid is formed, the nucleotides come either from the unlabeled nucleotides of the food or from the anemone nucleic acid pool.

GENERAL DISCUSSION

Some insight into the metabolic fate of food ingested by *Hydra* was obtained through studies in which *Hydra* was fed ^{35}S-labeled mouse tissue (Lenhoff 1961). Evidence was presented that (*a*) the *Hydra* cells take up 80 percent of the labeled tissue within 6 hours after food is swallowed; (*b*) no significant increase in labeled amino acids or peptides occurs in the lumen of the coelenteron during that 6 hours; (*c*) as the radioactivity of the protein within the gastrovascular cavity decreases, the radioactivity of the protein fraction of the *Hydra* cells increases; (*d*) a significant portion of the radioactivity from the ingested tissue is transferred from the highly radioactive endoderm to the ectoderm at about 24 hours following ingestion of the labeled food; and (*e*) the animals lose 25 percent of their radioactivity at a linear rate during the 5 days following feeding.

The most significant features of the present study showed (*a*) chemical changes in the food taking place within the coelenterate cells and (*b*) the rate of synthesis of coelenterate material from the food. These features are demonstrated by the relative increases and decreases of label in the alcohol-soluble fraction.

Alcohol-soluble proteins are usually not present in large amounts in animal tissues. The property of being soluble in 80 percent alcohol is usually ascribed to the so-called storage proteins (prolamines and zein) of seeds, or to a low-molecular-weight protein (insulin) from animals. Whether or not the alcohol-soluble material which we observed in mouse tissue, *Hydra*, or anemone represents a specific class of native proteins or consists instead of degradation products of larger alcohol-insoluble proteins cannot be determined from our studies. Because mouse tissues labeled by the methods described here do not always show a large amount of label in the Alc-S material, we assume for now that the large amount of label in the Alc-S fraction of the food represents partial protein degradation occurring in the dead tissue.

The increase of label in the Alc-S fraction observed both in *Hydra* and *Aiptasia* immediately following ingestion of labeled food is interpreted to be not an indication of a specific class of proteins synthesized on polysomes during this period but, rather, to represent metabolic intermediates of protein digestion. With *Hydra*, the formation of a labeled Alc-S fraction was slow enough for the experiments to show that this fraction is formed inside the gastrodermal cells. In the sea anemones, I was not able to show the kinetics of the increase of label in the Alc-S, although the great increase is evident (Figure 2*a*). There remains the possibility, therefore, that in *Aiptasia* the labeled Alc-S fraction is formed extracellularly before the food is phagocytized.

The experiments with both *Hydra* and *Aiptasia* suggest that the bulk of the hydrolysis of the ingested protein to amino acid takes place after the food is within the gastrodermal cells. The fractionation data described in this paper, however, still do not eliminate the possibility that free labeled amino acids are first released in the lumen and then are taken up immediately by the gastrodermis, where the amino acids are synthesized into the labeled Alc-S and Alc-I material, which we were able to measure there. This possibility seems unlikely in view of the data presented in this paper and in a previous study (Lenhoff 1961), and in veiw of the preponderance of cytological data showing phagocytosis of food particles.

Comparison of the results from *Hydra* with those from *Aiptasia* shows one particularly striking difference. In *Hydra* the proportion of label in the Alc-S fraction doubled in 7 hours and then remained at a nearly constant level for 1½ days before dropping off, whereas in *Aiptasia* the proportion of label in the Alc-S fraction increased sharply within 2–3 hours and then immediately began to disappear at a gradual rate.

These different patterns of Alc-S retention and disappearance may be simply due to such factors as (1) the different type of tissue fed to the animals; or (2) the different gastrodermal anatomies of the two animals. On the other hand, these differences in the Alc-S patterns may reflect (3) the presence and activities of the endosymbiotic zooxanthellae in *Aiptasia*. Such a possible influence of the algae might be investigated using symbiotic and aposymbiotic specimens of either *Aiptasia* or *Chlorohydra viridissima*.

SUMMARY

1. *Hydra littoralis* hydrolyzed much of the ingested ^{35}S-labeled food protein to a form insoluble in 5 percent trichloroacetic acid but soluble in 80 percent ethanol. One and one-half days later, about 50 percent of the label had disappeared from the alcohol-soluble fraction and had appeared in the alcohol-insoluble protein fraction.

2. *Aiptasia* fed ^{14}C-labeled protein differed from *Hydra littoralis* in that the label in the alcohol-soluble fraction in *Aiptasia* reached a peak 3 hours after ingestion of labeled food, and labeled alcohol-insoluble protein began to appear within 2 hours.

3. Discussed are (*a*) the rate of synthesis of coelenterate protein from digested labeled protein; (*b*) alcohol-soluble proteins in coelenterate; and (*c*) the failure of *Aiptasia* to incorporate significant label from digested ^{14}C-labeled protein into nucleic acids.

LITERATURE CITED

Gauthier, G. F. 1963. Cytological studies on the gastroderm of hydra. *J. Experimental Zoology* **152**: 13–39.

Lenhoff, H. M. 1961. Digestion of protein in *Hydra* as studied using radioautography and fractionation by differential solubilities. *Experimental Cell Research* **23**: 335–353.

———. 1968. Chemical perspectives on the feeding response, digestion, and nutrition of selected coelenterates. In *Chemical zoology*, vol. 2, M. Florkin and B. Scheer, eds., pp. 157–221. New York: Academic Press.

Loomis, W. F., and H. M. Lenhoff. 1956. Growth and sexual differentiation of hydra in mass culture. *J. Experimental Zoology* **132**: 555–573.

Roberts, R. B., P. H. Abelson, D. B. Cowie, E. T. Bolton, and R. Britten. 1955. *Studies of biosynthesis in Escherichia coli.* Publication 607. Carnegie Institution of Washington, D. C.

JOHN M. GOSLINE
Duke University, Durham, North Carolina

CHAPTER 14

Kinetics of Incorporation of ^{14}C-Proline into Mesogleal Protocollagen and Collagen of the Sea Anemone *Aiptasia*

Mesoglea, the layer between the epidermis and gastrodermis of coelenterates, can be thin (as the mesolamella in hydras), a thick mass of watery gel (as in "jellyfish"), or thick, cellular, and densely fibrous (as in sea anemones).

X-ray diffraction studies (Astbury 1940; Marks et al. 1949) indicated the presence of a collagenlike protein in coelenterate mesoglea. Electron micrographs by Grimstone et al. (1958) showed the mesoglea of the sea anemone *Metridium senile* to be a dense, cross-fibrillar structure. The fibrils showed regular banding with the period ranging from 220 to 250 angstroms, suggesting a collagen. Amino-acid analysis, X-ray diffraction and electron microscopy of mesoglea from *Metridium* and the Portuguese man-of-war *Physalia physalis* by Piez and Gross (1959) also indicated the presence of collagen.

In this chapter I show that, by taking advantage of two unique aspects of collagen, it is possible to investigate the kinetics of mesogleal collagen formation in the sea anemone *Aiptasia*. I refer to the well-known property of collagen to solubilize at high temperatures, forming gelatin, and to the formation of collagen hydroxyproline from the hydroxylation of peptide proline. It was feasible, therefore, to administer ^{14}C-proline to a sea anemone, to fractionate the collagen from the rest of the protein by autoclaving, and to measure new collagen formation by determining the amount of radioactive hydroxyproline in the anemone gelatin.

MATERIALS AND METHODS

Anemones of *Aiptasia* sp. were collected from Kaneohe Bay and cultured in the laboratory in plastic Petri dishes in about 40 ml of seawater. The animals were fed once a day on *Artemia* nauplii and the

---

An earlier version of this chapter appeared as "Kinetics of incorporation of proline–^{14}C into mesogleal protocollagen and collagen of the sea anemone, *Aiptasia*" by John M. Gosline and Howard M. Lenhoff, in *Comparative Biochemistry and Physiology* **26** (3): 1031–1039.

water was changed daily. Only small animals (oral disc diameter when expanded 2-3 mm) which had been kept in the laboratory for at least 1 week were used.

Incubations with ^{14}C-proline were carried out in the Petri dishes in which the animals were being maintained. There were usually two animals per dish. The animals, which were not fed the day prior to experiments, were given *Artemia* nauplii about 5 minutes before the ^{14}C-proline was added. About 2.5 μc of ^{14}C-proline in 10 μl of distilled water was injected into the coelenteron of each animal. After 1 hour the coelenteron was flushed several times with fresh seawater, and the surrounding seawater was changed.

The ^{14}C-proline (New England Nuclear Corporation, Boston, Mass.) had a specific activity of 206 mc/mM, and was reported to contain less than 0.05 percent hydroxyproline.

The following fractionation procedure was used. The animal was homogenized in a small volume of distilled water in a micro homogenizer, and the volume of the homogenate brought to 1.0 ml. A 0.1-ml sample of the homogenate was removed for assay. To the remaining 0.9 ml, 4.8 ml of hot (50° C) 95 percent ethanol was added to bring the volume to 5.7 ml and the final ethanol concentration to about 80 percent. When several animals were to be sampled over a period of time, the fractionation was stopped at this point and the homogenate kept in the freezer until all of the fractionations could be run at the same time.

The 80 percent ethanol suspension, either freshly prepared or removed from the freezer, was placed in a 45° C water bath and stirred every 10 minutes. After 30 minutes the suspension was centrifuged at 18,000 × g for 30 minutes and the supernatant passed through a Millipore filter (25-mm diameter, pore size 0.45 μ). The filtering removed any small particles not brought down by the centrifugation. The pellet was washed three times by resuspending in 5.0 ml of distilled water, centrifuging at 18,000 × g for 30 minutes, and removing the supernatant. (The particles in the supernatants from the initial centrifugation and from the washings were saved and counted. The radioactivity in the particles was figured into the final calculations.) The pellet was resuspended in 5.0 ml of distilled water and autoclaved for 12-15 hours at 15-18 psi. The resultant suspension was passed through a Millipore filter and the filter rinsed with distilled water. A 0.4-ml sample of the supernatant and the filter with the autoclave-insoluble material were put on different planchets.

This procedure yielded three major fractions. Fraction 1, the alcohol-soluble material plus wash (alcohol-soluble fraction) contained salts, small molecules, and alcohol-soluble lipid material. Fraction 2, the alcohol-insoluble, autoclave-soluble fraction (autoclave-soluble fraction) contained primarily gelatin from the mesoglea collagen which was solubilized during autoclaving. Fraction 3, the alcohol-insoluble, autoclave-insoluble fraction (autoclave-insoluble fraction) contained nucleic acids and proteins other than mesogleal collagen.

Planchets were counted on a Nuclear-Chicago model 470 gas flow counter with window. The activity in each fraction was expressed as percent of total activity. In most cases the radioactivity in recovered fractions was close to that of the calculated amount in the tissues. Of the 18 fractionations carried out, the mean recovery was 90.7 percent. None was less than 71 percent; only 2 were somewhat greater than 100 percent; and 11 ranged from 87 to 99.9 percent. An average of 382,000 counts/min was recovered from all animals.

Extracts of autoclaved tissues to be used for chromatography and amino acid analysis were hydrolyzed in 6 N HCl at 105° C for 12–15 hours in sealed tubes. The chromatograms were run in two dimensions (descending) on Whatman no. 4 filter paper using the solvent system of Bassham and Calvin (1957). A dye placed at the origin provided reference points on the chromatogram. Amino acid spots were developed with a spray of 0.25 percent ninhydrin in acetone.

The ^{14}C-hydroxyproline and ^{14}C-proline activity of the autoclave-soluble fraction was determined by making a one-dimensional chromatogram (phenol : water, 100 : 39 w/v) of the hydrolysate of this fraction. Pieces of the chromatogram 2 cm square were cut out, put on planchets, and counted. The activity in each piece was plotted against the distance along a one-dimensional chromatogram (Figure 1), and the percentages of radioactivity on the chromatogram as hydroxylproline and as proline were calculated. These figures, multiplied by the percentage of total activity in the autoclave-soluble fraction, gave the percentage of total activity as newly synthesized mesogleal collagen ^{14}C-hydroxyproline and ^{14}C-proline.

A quantitative colorimetric assay for hydroxyproline was carried out after the method of Woessner (1961) as modified by Lenhoff and Bovaird (1961). Proline was assayed by the method of Wren and Wiggall (1965). The results are expressed in micrograms hydroxyproline per milligram dry weight of animal tissues, and as the ratio of proline to hydroxyproline.

RESULTS

Demonstration of Hydroxyproline in the Hydrolysate of Autoclave-Soluble Material

To determine whether the sea anemone *Aiptasia* has a collagen that could be solubilized by autoclaving, hydrolysates of autoclave-soluble material were analyzed for hydroxyproline. After being sprayed with ninhydrin, a two-dimensional chromatogram of a hydrolysate of such a fraction from a large *Aiptasia* showed a prominent yellow spot in the area having R_F values similar to pure hydroxyproline. A similar distribution of

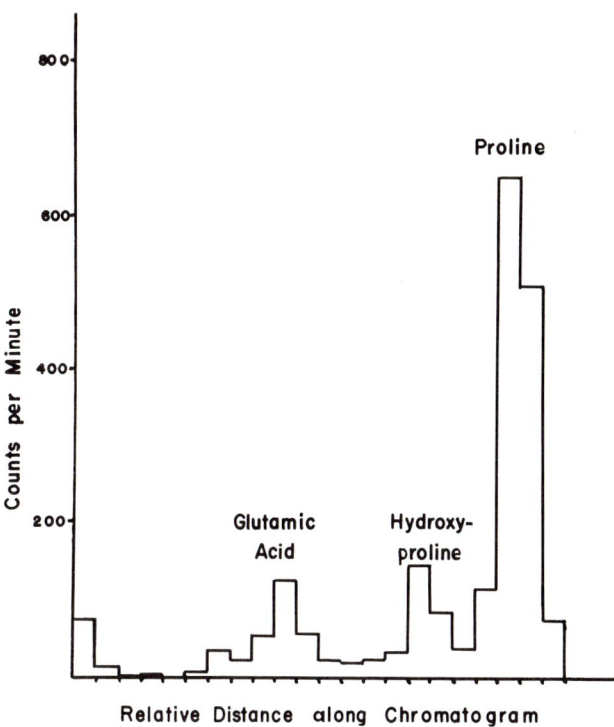

FIG. 1. Distribution of radioactivity along a one-dimensional chromatogram of a hydrolysate of the autoclave-soluble fraction of *Aiptasia* sp. given 1 μc ^{14}C-proline 24 hours prior to fractionation.

ninhydrin spots was observed in a chromatogram of a hydrolysate of a rat tail tendon and of a hydrolysate of the autoclave-soluble fraction of pieces of the body wall of another sea anemone, *Marcranthea cookei*.

A quantitative hydroxyproline assay (Lenhoff and Bovaird 1961) of the hydrolysate of the autoclave-soluble fraction from three small anemones gave an average of 2.3 μg hydroxyproline per animal. The hydroxyproline content of the mesogleal collagen of *Metridium* as reported by Piez and Gross (1959) is 4.98 percent by weight. Assuming that *Aiptasia* mesogleal collagen is in the same range, 2.3 μg of hydroxyproline represents about 0.05 mg of collagen. Since the dry weights of animals approximately the same size ranged from 0.75 to 1.0 mg, the mesogleal collagen can be considered to make up about 5 percent of the dry weight of the animal and, accordingly, about 10 percent of the protein of the anemone. Many of these figures are approximations, but they do indicate that the hydroxyproline values are within the expected range.

Distribution of Radioactivity into Three Fractions Obtained from *Aiptasia*

Animals were given a 1-hour pulse of ^{14}C-proline, the coelenterons were flushed, and the animals were fractionated at intervals from 1½ to 48 hours following the administration of the label. (The points in Figures 2-6 marked with triangles indicate data from four experiments in which the coelenterons were not flushed.)

The kinetics of the distribution of radioactivity clearly indicate that as the alcohol-soluble fraction disappeared (Figure 2), both the autoclave-insoluble fraction (Figure 3) and autoclave-soluble fraction (Figure 4) increased. By 10 hours following administration of the ^{14}C-proline, the bulk of the conversion had taken place. By 48 hours the alcohol-soluble fraction leveled at about 20 percent of the radioactivity in the animal, the autoclave-insoluble fraction contained nearly 60 percent of the activity, and the autoclave-soluble fraction contained the remaining 20 percent.

Semilogarithmic plots of the data over the first 8 hours showed that the alcohol-soluble fraction lost about 12 percent of its radioactivity per hour, while the autoclave-insoluble fraction and autoclave-soluble fraction increased in radioactivity at about 6.5 and 4 percent per hour, respectively. That the alcohol-soluble fraction disappeared at a rate slightly faster than the other two fractions increased might be explained by leakage of label from the animal or by the oxidative metabolism of proline or its by-products.

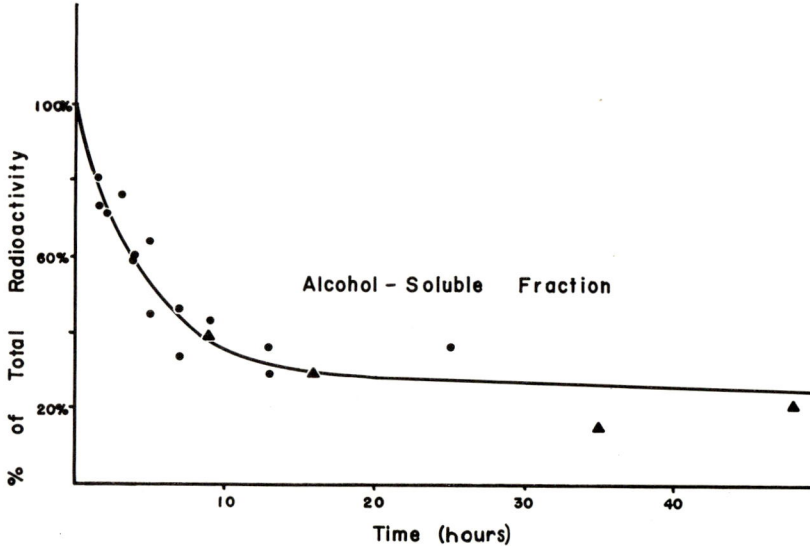

FIG. 2. Changes in the percentages of total radioactivity in the alcohol-soluble fraction of *Aiptasia* sp. given 2.5 μc ^{14}C-proline at time zero.

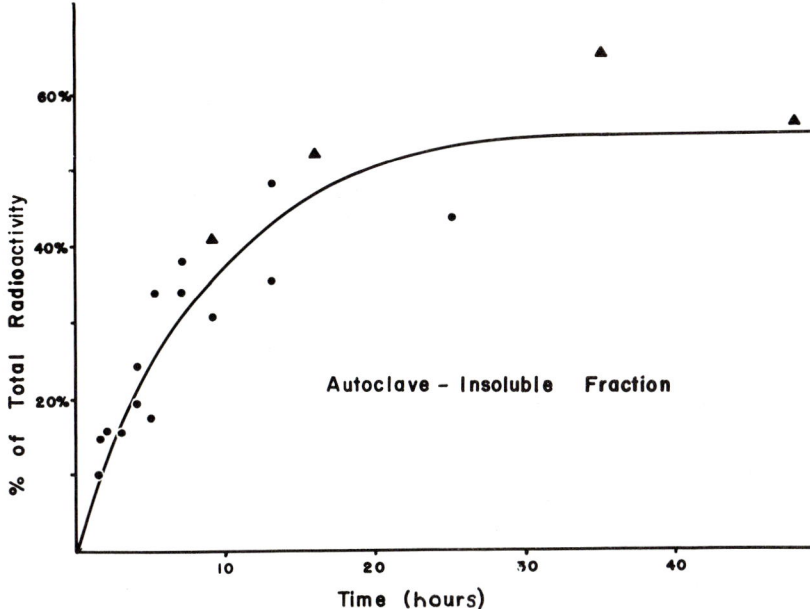

FIG. 3. Changes in the percentages of total radioactivity in the autoclave-insoluble fraction of *Aiptasia* sp. given 2.5 μc ^{14}C-proline at time zero.

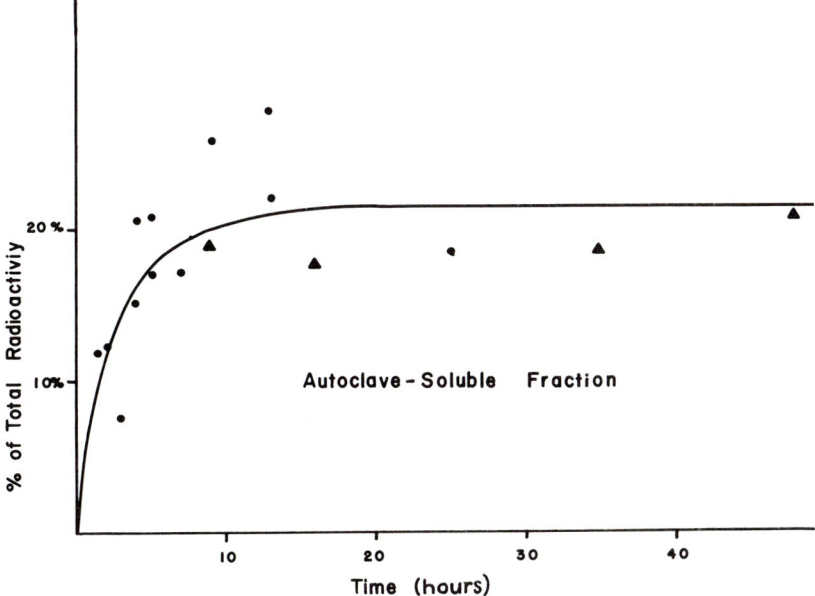

FIG. 4. Changes in the percentages of total radioactivity in the autoclave-soluble fraction of *Aiptasia* sp. given 2.5 μc ^{14}C-proline at time zero.

Evidence for Protocollagen from Analyses of Autoclave-Soluble Fractions

Analyses of the distribution of radioactivity on one-dimensional chromatograms of acid hydrolysates of various autoclave-soluble fractions showed, as seen in the histogram in Figure 1, that there are three major peaks of radioactivity. The proline and hydroxyproline peaks were identified by their positions relative to the marker dye; the third peak was tentatively considered to be glutamic acid, a major breakdown product of proline.

Figure 1 shows the distribution of radioactivity in the autoclave-soluble fraction of an animal fractionated 24 hours after the administration of ^{14}C-proline. The amount of radioactivity in the proline spot is roughly four to five times that in the hydroxyproline spot. This disproportionate distribution of radioactivity was found to be even more pronounced when the early alcohol-soluble fractions were analyzed. It is doubtful that the high proline content was contamination from the alcohol-soluble fraction because the alcohol-insoluble pellet was washed three times in distilled water. Examination of Figure 5, which is a plot of the ^{14}C-proline:^{14}C-hydroxyproline ratio of the autoclave-soluble fractions, shows that this ratio was about 20:1 during the first 2 hours following administration of ^{14}C-proline, and then leveled off quickly to about 1:1 by 4 hours. The proline:hydroxyproline ratio of the autoclave-soluble fraction of unlabeled *Aiptasia* from Hawaii was about 4:1, whereas that of Florida *Aiptasia* was about 6.7:1. Despite the differences in these figures, the

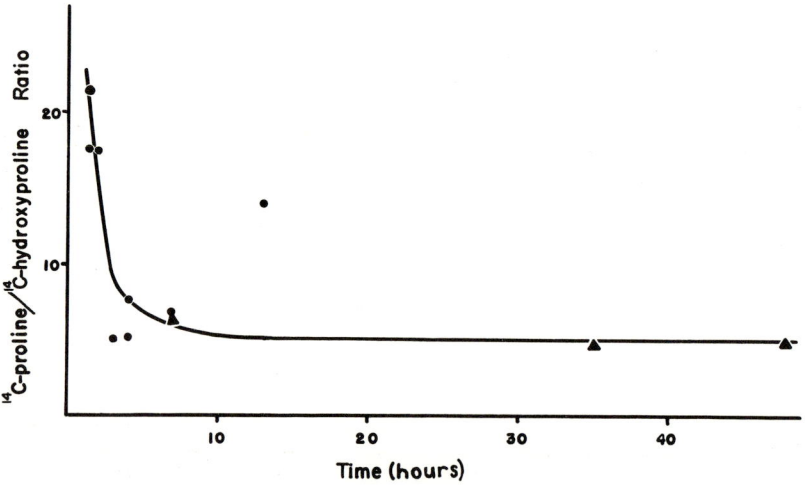

FIG. 5. Changes in the ^{14}C-proline to ^{14}C-hydroxyproline ratio of the autoclave-soluble fraction of *Aiptasia* sp. given ^{14}C-proline at time zero.

Incorporation of ^{14}C-Proline by *Aiptasia*

ratios approximate those found with the radioactive proline and hydroxyproline in Hawaiian anemones 36–48 hours after the administration of ^{14}C-proline. The possible significance of the high and changing ^{14}C-proline:^{14}C-hydroxyproline ratio (Figure 5) is dealt with under "Discussion."

Evidence that there was no significant carry-over of free ^{14}C-proline from the alcohol-soluble fraction into the autoclave-soluble material (protocollagen) analyzed in Figures 1 and 5 was obtained by analyzing the radioactivity in the three washes of each alcohol-insoluble pellet. Typically, wash no. 1 contained about 5 percent of the radioactivity in the total alcohol-soluble fraction. Wash no. 2 contained about 0.5 percent of the radioactivity; and wash no. 3, about 0.2 percent. Even if the carry-over were assumed to be five times the radioactivity in wash no. 3, and the ratio of ^{14}C-proline to ^{14}C-hydroxyproline was recalculated, the relationship shown in Figure 5 still holds.

Because the actual hydroxylation of peptide proline to hydroxyproline is the near-final step in the biosynthesis of collagen, I considered that the clearest indication of the rate and proportion of collagen synthesized would come from analysis of the increase in radioactive hydroxyproline in the autoclave-soluble fraction. Figure 6 shows a plot of the increase in the percentage of radioactivity as hydroxyproline. The curve shows that approximately 2–3 percent of the ^{14}C-proline administered is converted to hydroxyproline, and that the rate of this conversion begins to level off

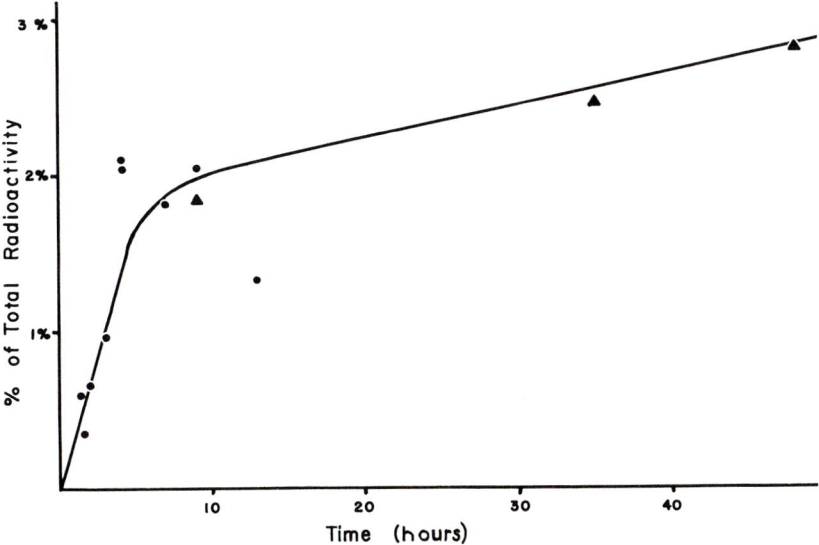

FIG. 6. Changes in the percentages of total radioactivity as ^{14}C-hydroxyproline in the autoclave-soluble fraction of *Aiptasia* sp. given 2.5 μc ^{14}C-proline at time zero.

within 9 hours. The data are not sufficiently accurate to allow an estimation of the initial rate of hydroxyproline formation. These analyses suggest that the rate of hydroxyproline formation is considerably less than the rate of incorporation of label into either the autoclave-soluble or the autoclave-insoluble fractions.

DISCUSSION

In these experiments we measure the rate of the *in vivo* synthesis of mesogleal collagen in the sea anemone *Aiptasia* sp. The methods consisted of following (*a*) the rate of incorporation of ^{14}C-proline into a fraction insoluble in 80 percent ethanol, but soluble on autoclaving, and (*b*) the rate of conversion of ^{14}C-proline into ^{14}C-hydroxyproline.

The mere measurement of the total ^{14}C-hydroxyproline formed in the whole animal could not be used as a criterion for mesogleal collagen synthesis, because hydroxyproline also occurs as a component of the disulfide-linked proteins forming the major part of the capsule of *Aiptasia* nematocysts (Blanquet and Lenhoff 1966). Nematocyst capsules, however, are insoluble on autoclaving (Lenhoff et al. 1957; Mariscal and Lenhoff, unpublished), and, hence, did not contaminate the autoclave-soluble mesoglea fraction with nematocyst hydroxyproline-rich proteins.

The results show that the alcohol-insoluble fraction of *Aiptasia*, like the body wall of *Metridium* and the float of *Physalia*, form on autoclaving a gelatinlike substance. This material contains both proline and hydroxyproline in a ratio of about 6:1. This ratio in the anemone *Metridium* was found to be about 1:1 (Piez and Gross 1959). Other coelenterate mesogleas have been analyzed for hydroxyproline; the mesoglea from the hydromedusa *Aequorea aequorea* contains 2–3 percent hydroxyproline (Blanquet and Lenhoff 1966), and mesoglea from the scyphomedusa *Pelagia noctiluca* contains 3.6 percent hydroxyproline (Chapman 1959).

We show that free proline is the source of both bound proline and bound hydroxyproline in the mesogleal collagen of coelenterates. This has previously been shown to be true in vertebrate collagens (Kivirikko and Prockop 1967) and earthworm-cuticle collagen (Fujimoto and Adams 1965). In addition, the data in Figure 5 indicate that the autoclave-soluble fractions isolated during the first few hours after ^{14}C-proline administration contain an unhydroxylated collagen precursor, probably a type of protocollagen. The existence of such a collagen intermediate in *Aiptasia* is in accord with current views of collagen biosynthesis (see Kivirikko and Prockop 1967; Udenfriend 1966). Proteins having greater-than-normal proline:hydroxyproline ratios and which are presumably collagen precursors have been found in vertebrates (Udenfriend 1966) and in earthworms (Fujimoto and Adams 1965). I now report the presence of such material in a sea anemone.

The high proline:hydroxyproline ratio of the autoclave-soluble fraction of unlabeled animals can be explained in several ways; (1) contamination from the alcohol-soluble fraction; (2) the presence of other, noncollagen proteins; or (3) an unusually high proline:hydroxyproline ratio in *Aiptasia* collagen.

A good measure of the synthesis of the final collagen molecule may be the amount of increase in ^{14}C-hydroxyproline in the autoclave-soluble fraction. When a ^{14}C-proline:^{14}C-hydroxyproline ratio is reached approximating that of the unlabeled mesogleal collagen, the hydroxylation process appears complete. Examination of Figures 5 and 6 shows that most of the hydroxyproline synthesis had taken place after 8 hours, and that the ^{14}C-proline:^{14}C-hydroxyproline ratio had reached a constant value.

Mesoglea can be considered as a prototype of epithelial basement membranes, and as a primitive connective tissue. Little is known about its chemical composition, factors influencing its biosynthesis, its physiological role, and the cellular site of its origin. Extension of the approach described in this paper may answer some of these questions, as well as provide general information on collagen biosynthesis.

SUMMARY

1. Methods are described for labeling and purifying mesogleal collagen from the sea anemone *Aiptasia* sp. These techniques were applied to the study of the kinetics of the *in vivo* formation of mesogleal collagen.

2. Free ^{14}C-proline, injected into the coelenteron, was incorporated into bound ^{14}C-proline and bound ^{14}C-hydroxyproline of a protein(s) which becomes soluble on autoclaving.

3. The autoclave-soluble material, in addition to containing collagen, is thought to contain an unhydroxylated collagen precursor. Within a few hours after the ^{14}C-proline was administered, the ratio of ^{14}C-proline to ^{14}C-hydroxyproline was 20:1; the proline:hydroxyproline ratio of unlabeled material was about 6:1.

LITERATURE CITED

Astbury, W. T. 1940. The molecular structure of the fibers of the collagen group. *J., International Society of Leather Trades Chemists* **24**: 69–92.

Bassham, J. A., and M. Calvin. 1957. *The path of carbon in photosynthesis.* Englewood Cliffs, N.J.: Prentice-Hall.

Blanquet, R., and H. M. Lenhoff. 1966. A disulfide-linked collagenous protein of nematocyst capsules. *Science* **154**: 152–153.

Chapman, G. 1959. The mesoglea of *Palagia noctiluca. Quarterly J. of Microscopical Science* **100**: 599–610.

Grimstone, A. V., R. W. Horne, C. F. A. Pantin, and E. Robson. 1958. The fine structure of the mesenteries of the sea-anemone *Metridium senile. Quarterly J. of Microscopical Science* **99**: 523–540.

Fujimoto, D., and E. Adams. 1965. Proline incorporation into the proline and hydroxyproline of earthworm-cuticle collagen. *Biochimica et biophysica acta* **107**: 232–246.

Kivirikko, K. O., and D. J. Prockop. 1967. Hydroxylation of proline in synthetic polypeptides with purified protocollagen hydroxylase. *J. Biological Chemistry* **242**: 4007–4012.

Lenhoff, H. M., and J. Bovaird. 1961. A quantitative chemical approach to problems of nematocyst distribution and replacement in *Hydra. Developmental Biology* **3**: 227–240.

Lenhoff, H. M., E. Kline, and R. Hurley. 1957. A hydroxyproline-rich, intracellular collagen-like protein of *Hydra* nematocysts. *Biochimica et biophysica acta* **26**: 204–205.

Marks, M. H., R. S. Bear, and C. H. Black. 1949. X-ray diffraction evidence of collagen-like protein fibers in the Echinodermata, Coelenterata, and Porifera. *J. Experimental Zoology* **111**: 55–78.

Piez, K. A., and J. Gross. 1959. The amino acid composition and morphology of some invertebrate and vertebrate collagens. *Biochimica et biophysica acta* **34**: 24–39.

Udenfriend, S. 1966. Formation of hydroxyproline in collagen. *Science* **152**: 1335–1340.

Woessner, J. F., Jr. 1961. The determination of hydroxyproline in tissue and protein samples containing small proportions of this amino acid. *Archives of Biochemistry and Biophysics* **93**: 440-447.

Wren, J. J., and P. H. Wiggall. 1965. An improved colorimetric method for the determination of proline in the presence of other ninhydrin-positive compounds. *Biochemical J.* **94**: 216–220.

RICHARD N. MARISCAL
Florida State University, Tallahassee

CHAPTER 15

Effect of a Disulfide Reducing Agent on the Nematocyst Capsules from Some Coelenterates, with an Illustrated Key to Nematocyst Classification

The protein composition of coelenterate nematocyst capsules has only recently been clarified. The earlier works of Brown (1950), Hamon (1955a,b), and Yanagita (1959) demonstrated the presence of disulfide linkages in the capsule wall of nematocysts. On the other hand, Lenhoff, Kline, and Hurley (1957) and Lenhoff and Kline (1958) reported that the nematocyst capsules of *Hydra littoralis* and *Physalia* are composed of protein containing large quantities of hydroxyproline. Because the presence in protein of the imino acid hydroxyproline is diagnostic of collagens, and because collagens were thought to be devoid of disulfide bonds, the above data seemed contradictory.

Blanquet and Lenhoff (1966), however, have recently presented data that reconcile the above reports. They found that the capsule of the acontial nematocysts of the anemone *Aiptasia pallida* contains a single hydroxyproline-rich, collagenlike protein linked by disulfide bonds. Their experiments showed that a number of disulfide reducing agents, including dithioerythritol (DTE), sodium thioglycolate, cysteine, and mercaptoethanol are all effective in dissolving nematocyst capsules from *Aiptasia*. Because DTE, acting at low concentration and slightly alkaline pH, was the most effective of the above reducing agents, we tested the action of this compound on the nematocysts of a number of other coelenterates to determine if they, too, contained disulfide bonds and could be dissolved.

MATERIALS AND METHODS

The coelenterates used were collected in Kaneohe Bay. Five species of corals, four sea anemones, one zoanthid, one scyphozoan, and one hydroid were examined.

Tentacles were removed either with iridectomy scissors or watchmaker's forceps and were placed on a clean glass slide in a drop of seawater.

An earlier version of this chapter appeared as "Effect of a disulfide reducing agent on coelenterate nematocyst capsules" by R. N. Mariscal and H. M. Lenhoff, in *Experientia* **25**: 330–331.

A cover slip was added and a uniform smear was prepared by pressing on the cover slip with a foil-covered rubber stopper. This method not only freed nematocysts from most of the surrounding tissue and mucus, but also caused some to discharge.

The various nematocyst types were identified by a key (see Appendix) prepared from the following sources: Weill 1934; Carlgren 1940, 1945; Cutress 1955; Hand 1961; Mackie and Mackie 1963; Deroux 1966; Bouillon and Deroux 1967; and Werner 1965.

Three drops of 10^{-2} M DTE made up in tris-(hydroxy)-methylaminomethane buffer at pH 9.1 were added to the preparation and the cover slip was replaced. The time required for the nematocyst capsules to dissolve completely at 28° C was recorded.

The above technique, although seemingly simple, presented some problems. It proved difficult to obtain enough nematocysts free of the surrounding tissue-mucus smear so that the capsule was available to the DTE solution. Discharged nematocysts embedded in mucus were not affected by the DTE, even at high pH. Thus it was necessary to include in the results only those nematocysts which had been observed to be free of adherent tissue or mucus, and which had been bathed in the DTE solution directly.

In general, the nematocysts had to be discharged before the DTE was able to dissolve the capsules. Most nematocysts that were either undischarged or partially discharged so that the shaft was blocking the opercular opening did not respond to the DTE, even after several hours' exposure.

RESULTS

The first detectable effect of the DTE on the discharged nematocyst capsule was a light buckling of the normally smooth capsule wall. The capsule wall then appeared to become less dense, as observed microscopically. Solubilization appeared to start from the inner surface of the capsule wall and to proceed outward. Before the capsule finally dissolved, a thin membranelike coat remained visible, but it too eventually disappeared. The everted shaft, thread, and barbs did not dissolve during the period of observation.

As will be noted in Table 1, most of the nematocysts examined from all three classes of coelenterates dissolved in 1–3 minutes, if they were going to do so at all. For example, the stenoteles of the single hydroid tested (*Pennaria tiarella*), the nematocysts of the scyphozoan, the zoanthid, and all the anemones dissolved in 1–3 minutes. A solubilization time greater than 3 minutes for these groups generally indicated that the nematocysts were not going to be affected by the DTE under these conditions.

The nematocysts of the corals seemed the most resistant to dissolution by DTE. Of the four corals tested, only the microbasic mastigophore of *Porites compressa* dissolved within 3 minutes, whereas those of *Cyphastrea* required 9-10 minutes to dissolve. All other coral nematocysts examined remained undissolved after an exposure of 10 to 30 minutes to DTE. Interestingly, the spirocysts were not affected by the DTE.

As indicated under "Methods," the above data are based on discharged nematocyst capsules only. The unfired nematocysts of the various coelenterates tested, with one exception, remained unaffected by the DTE in this study. The single exception was the microbasic amastigophore of the swimming sea anemone *Boloceroides*; both the discharged and undischarged amastigophores dissolved in 1-2 minutes.

DISCUSSION

The present study extends the findings of Brown (1950), Yanagita (1959), and Blanquet and Lenhoff (1966) from work done mostly on anemone nematocysts, and shows that the nematocyst capsules from a wide variety of coelenterates can be dissolved by a disulfide reducing agent. Methods for purifying large numbers of nematocysts from other coelenterates are needed, however, to determine the complete chemical nature of the capsule walls. Recently, Fishman and Levy (1967) extended the findings of Blanquet and Lenhoff (1966), and showed that nematocysts isolated from *Metridium marginatum* are dissolved by DTE and contain hydroxyproline.

That some of the nematocysts examined required 9-10 minutes to dissolve, or did not dissolve at all during the period under examination, remains enigmatic. This is especially true for the coral nematocysts. Perhaps the relatively large amounts of mucus liberated by corals (compared to the small amount given off by the other coelenterates examined) protects the nematocysts from dissolution by DTE. On the other hand, some coral nematocyst capsules may differ chemically from the others investigated to date.

Except for the nematocysts from *Boloceroides*, the other nematocysts examined did not dissolve in DTE unless they had been completely discharged. Yanagita (1959) found that partially discharged acontial nematocysts of the anemone *Diadumene* are unaffected by 0.125 M sodium thioglycolate at pH 8.6, whereas the discharged nematocysts dissolve shortly after application of the thioglycolate. In the present study, with the single exception noted, both undischarged and partially discharged nematocysts (those with the shaft partially everted and plugging the opercular opening) appeared to be totally unaffected by the DTE. Blanquet and Lenhoff (1966) observed that dissolution of the

TABLE 1 Effect of dithioerythritol (DTE) on nematocyst capsules and spirocysts

Species	Nematocyst Type	Time for Capsule to Dissolve (minutes)	Comments
Class Hydrozoa			
1. *Pennaria tiarella*	Large stenotele	2	
	Small stenotele	2	
Class Scyphozoa			
1. *Cassiopeia* sp.	Holotrichous isorhizas	—	Undissolved after 18 minutes
	Microbasic homotrichous eurytele	1	
Class Anthozoa			
Order Actiniaria (anemones)			
1. *Boloceroides lilae*	Spirocysts	—	Undissolved after 30 minutes
	Microbasic amastigophores (discharged and undischarged)	1–2	

2. *Macranthea cookei*	Microbasic-p-mastigophores	3	
3. *Anthopleura* sp.	Microbasic amastigophore	2	
4. *Aiptasia pulchella*	Microbasic amastigophore (acontia and tentacles)	1–2	
	Spirocysts	–	Undissolved after 30 minutes
Order Zoanthidea			
1. *Zoanthus sandwichensis*	Holotrichous isorhizas	2	
Order Schleractinia (Madreporaria) (stony corals)			
1. *Cyphastrea ocellina*	Holotrichous isorhiza	9	
	Microbasic mastigophore	10	
2. *Porites compressa*	Microbasic mastigophore	3	
3. *Tubastrea manni*	Microbasic-b-mastigophore	–	Undissolved after 10 minutes
4. *Pocillopora meandrina*	Microbasic mastigophore	–	Partially dissolved after 15 minutes (?)
5. *Fungia scutaria*	Atrichous isorhiza	–	Undissolved after 15 minutes
	Microbasic mastigophore	–	Undissolved after 30 minutes

everted thread of *Aiptasia* nematocysts, in contrast to the capsule, requires a relatively higher pH and longer periods of exposure to the reducing agents. They never were able to dissolve the uneverted threads nor the barbs of the everted ones, even at pH 11.

Spirocysts, as shown through histochemical and light microscopy techniques, differ from nematocysts (see Hyman 1940, for general discussion). The current study appears to indicate that spirocysts are chemically different in another way: they are not affected by the disulfide reducing agent DTE. The specific chemical composition of spirocysts is still unknown, however, and it will not be possible to examine them much further until methods are devised for purifying them in large quantities.

SUMMARY

1. The nematocyst capsules from a wide variety of Hawaiian coelenterates were treated with dithioerythritol (DTE), a reducing agent specific for disulfide bonds.
2. The solubilization of the discharged capsules was generally completed within 1–3 minutes.
3. The capsules of the spirocysts were unaffected by the DTE.
4. The coral nematocysts proved to be more resistant to DTE than did those from the other coelenterates.
5. These data show that the presence of disulfide linkages may be common to many coelenterate nematocysts and tend to support previous findings that the capsule wall of coelenterate nematocysts is composed of disulfide-linked collagens.

APPENDIX

This illustrated key to the classification of nematocysts is included in the hope that it will prove useful to English-speaking workers in the field of coelenterate biology. The best recent discussion of nematocyst classification is that of Werner (1965), and special thanks are due him for permission to use the figures from that paper. Figures 7 and 17 are after Mackie and Mackie (1963) and Deroux (1966). All the others are from Werner (1965). This latter paper should be consulted for a detailed discussion concerning the source of the figures and distribution of the various nematocyst types in the phylum. Thanks are also due Dr. Werner for many helpful comments and suggestions concerning nematocyst classification and morphology, as well as to Dr. Cadet Hand for many stimulating discussions concerning nematocysts, for reviewing the classification scheme, and for offering additional suggestions.

As more work is done with nematocysts, especially with the electron microscope, some of Weill's (1934) original classification scheme may have to be revised. For example, Hand (1961: 49) pointed out that some atrichous isorhizas, when viewed with the electron microscope, turn out to have small spines which cannot be resolved with the light microscope. Since it is obviously impractical for most coelenterate biologists to study each nematocyst in question under the electron microscope and in order to retain the basic usefulness of Weill's system, I have added the modifier "well-developed" to describe the condition of the spines on the thread of the various isorhizas. "Well-developed" here means spines which are readily visible under the high dry objective of the average laboratory microscope.

In addition, Werner (personal communication) suggested that both the homotrichous and heterotrichous microbasic euryteles can be grouped together simply as microbasic euryteles. Similarly, he feels that the three types of macrobasic euryteles would best be grouped under this heading because of overlapping features. Cutress (1955) and Hand (1961) covered additional difficulties in nematocyst classification, nearly all of which must await clarification by means of the electron microscope. Until that time, and because of its proven usefulness, Weill's basic classification system has been retained with the addition of the new nematocyst types discovered since his original publications.

Classification of Coelenterate Nematocysts*

I. ASTOMOCNIDES. Thread closed at the tip
    A. RHOPALONEMES. Thread club-shaped
        1. Anacrophores. Thread without an apical projection (1)[†]
        2. Acrophores. Thread with an apical projection (2)
    B. DESMONEMES. Thread forming a corkscrewlike coil (3)
II. STOMOCNIDES. Thread open at the tip
    A. HAPLONEMES. Thread without a well-defined shaft
        1. Isorhizas. Thread of the same diameter throughout (glutinants)
            a. Atrichous. Thread without well-developed spines (small glutinant) (4)
            b. Basitrichous. Thread with well-developed spines only at base (5)
            c. Merotrichous. Thread with well-developed spines on the intermediate portion only (6)

*After Weill 1934; Carlgren 1940, 1945; Cutress 1955; Hand 1961; Mackie and Mackie 1963; Werner 1965; and Deroux 1966. See illustration of macrobasic mastigophore (Figure 12) for basic nematocyst terminology.

[†]The numbers in parentheses following each description refer to the figures of the different nematocyst types.

d. Apotrichous. Thread with well-developed spines on the distal portion only (7)
e. Holotrichous. Thread with well-developed spines along whole length (large glutinant) (8)
2. Anisorhizas. Thread slightly dilated towards base.
   a. Homotrichous. Thread spiny throughout; spines all of equal size (9)
   b. Heterotrichous. Thread spiny throughout; spines larger at base of thread (10)

B. HETERONEMES. Thread with a well-defined shaft
  1. Rhabdoides. Shaft cylindrical, of the same diameter throughout
     a. Mastigophores. Thread continues beyond the shaft
        (1). Microbasic. Shaft short, less than 3 times capsule length (11)
           (a). Microbasic b-mastigophore. Shaft tapers gradually into thread
           (b). Microbasic p-mastigophore. Shaft tapers abruptly into thread; V-shaped notch prominent at base of unfired shaft
           (c). Microbasic q-mastigophore. Detachable dart on the end of the invaginated shaft
        (2). Macrobasic. Shaft long, more than 4 times capsule length (12)
     b. Amastigophores No thread beyond the shaft
        (1). Microbasic. Shaft short, less than 3 times capsule length (13)
        (2). Macrobasic. Shaft long, more than 4 times capsule length (14)
  2. Rhopaloides. Shaft of unequal diameter
     a. Euryteles. Shaft dilated distally
        (1). Microbasic. Shaft short, less than 3 times capsule length
           (a). Homotrichous. Spines of shaft all the same size (15)
           (b). Heterotrichous. Spines of shaft of unequal size (16)
           (c). Semiophoric. Thread bent whiplike with large flat spine medially (17)
        (2). Macrobasic. Shaft long, more than 4 times capsule length

(a). Telotrichous. Spines distal only on shaft (18)
(b). Merotrichous. Spines not distal, found only on shaft area of uniform diameter proximal to terminal swelling (19)
(c). Holotrichous. Shaft spiny along whole length (20)
    b. Stenoteles. Shaft dilated at base; 3 spines especially strongly developed (penetrants) (21)
  3. Birhopaloides. Shaft of unequal diameter at distal and proximal ends (22)
III. SPIROCYSTS. Thin, single-walled capsule containing a long, spirally coiled, dense-appearing thread of uniform diameter; no shaft or spines distinguishable (23)

LITERATURE CITED

Blanquet, R., and H. M. Lenhoff. 1966. A disulfide-linked collagenous protein of nematocyst capsules. *Science* **154**: 152–153.

Bouillon, J., and C. Deroux. 1967. Remarques sur les cnidaires du type de *Microhydrula pontica* Valkanov 1965, trouvés a Roscoff. *Cahiers de Biologie Marine* **8**: 253–272.

Brown, C. H. 1950. Keratins in invertebrates. *Nature* **166**: 439.

Carlgren, O. 1940. A contribution to the knowledge of the structure and distribution of the cnidae in the Anthozoa. *Lunds Universitets Årsskrift, N.F.* **36**: 1–62.

———. 1945. Further contributions to the knowledge of the cnidom in the Anthozoa especially in the Actiniaria. *Lunds Universitets Årsskrift, N.F.* **41**: 1–24.

Cutress, C. E. 1955. An interpretation of the structure and distribution of cnidae in Anthozoa. *Systematic Zoology* **4**: 120–137.

Deroux, G. 1966. Un nouveau type de Nématocystes rhopaloides (R. Weill, 1934): les Euryteles microbasiques (hétérotriches) Semiophores. *Compte rendu de l'Académie des sciences.* Paris **263**: 760–763.

Fishman, L., and M. Levy. 1967. Studies on the nematocyst capsule protein from the sea anemone *Metridium marginatum. Biological Bulletin, Woods Hole* **133**: 464–465.

Hamon, M. 1955a. Cytochemical research on coelenterate nematocysts. *Nature* **176**: 357.

———. 1955b. Recherches histochimiques sur les nématocystes de Coelentérés. *Bulletin, Société d'histoire naturelle de l'Afrique due Nord* (Algiers) **46**: 169–179.

Hand, C. 1961. Present state of nematocyst research: types, structure and function. In *The biology of hydra and of some other coelenterates: 1961*, H. M. Lenhoff and W. F. Loomis, eds., pp. 187–202. Coral Gables: University of Miami Press.

Hyman, L. 1940. *The invertebrates: Protozoa through Ctenophora*, vol. 1. New York: McGraw-Hill.

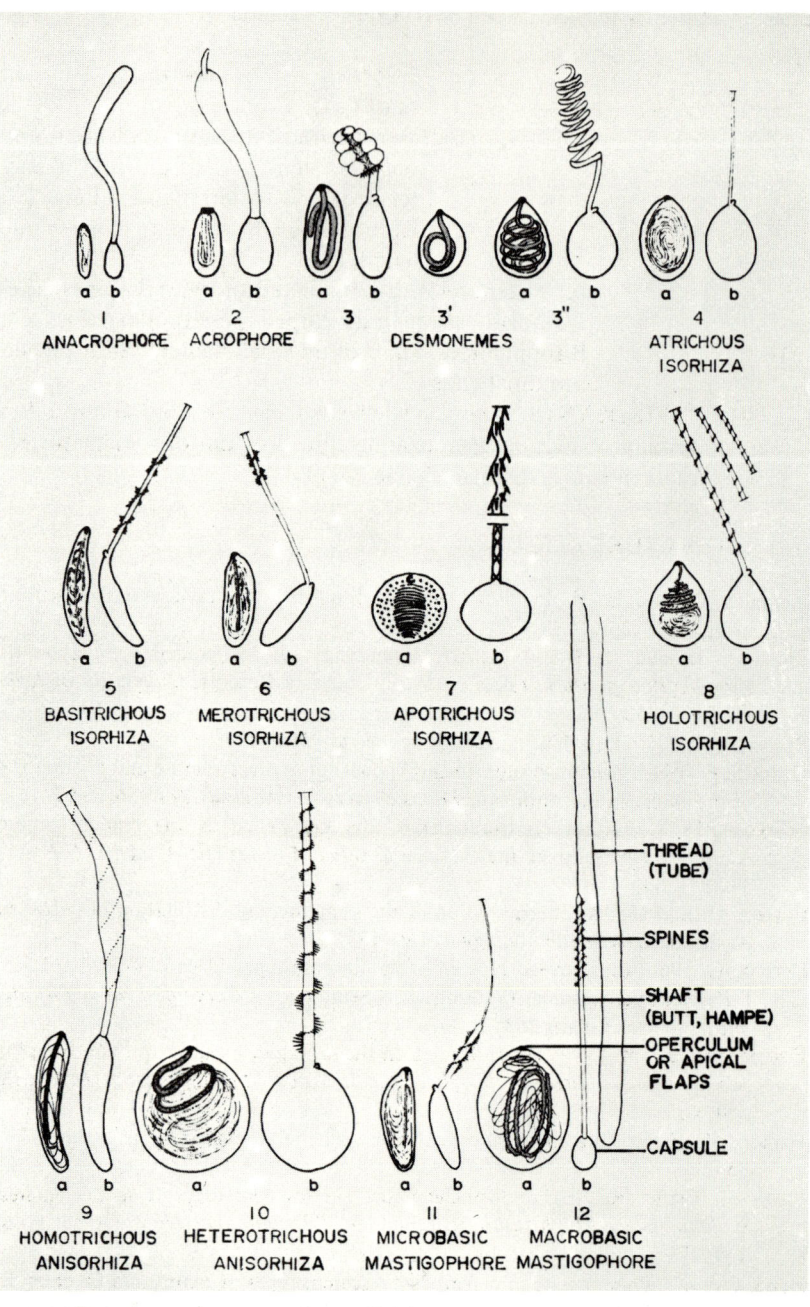

FIGS. 1–12. Coelenterate nematocysts.

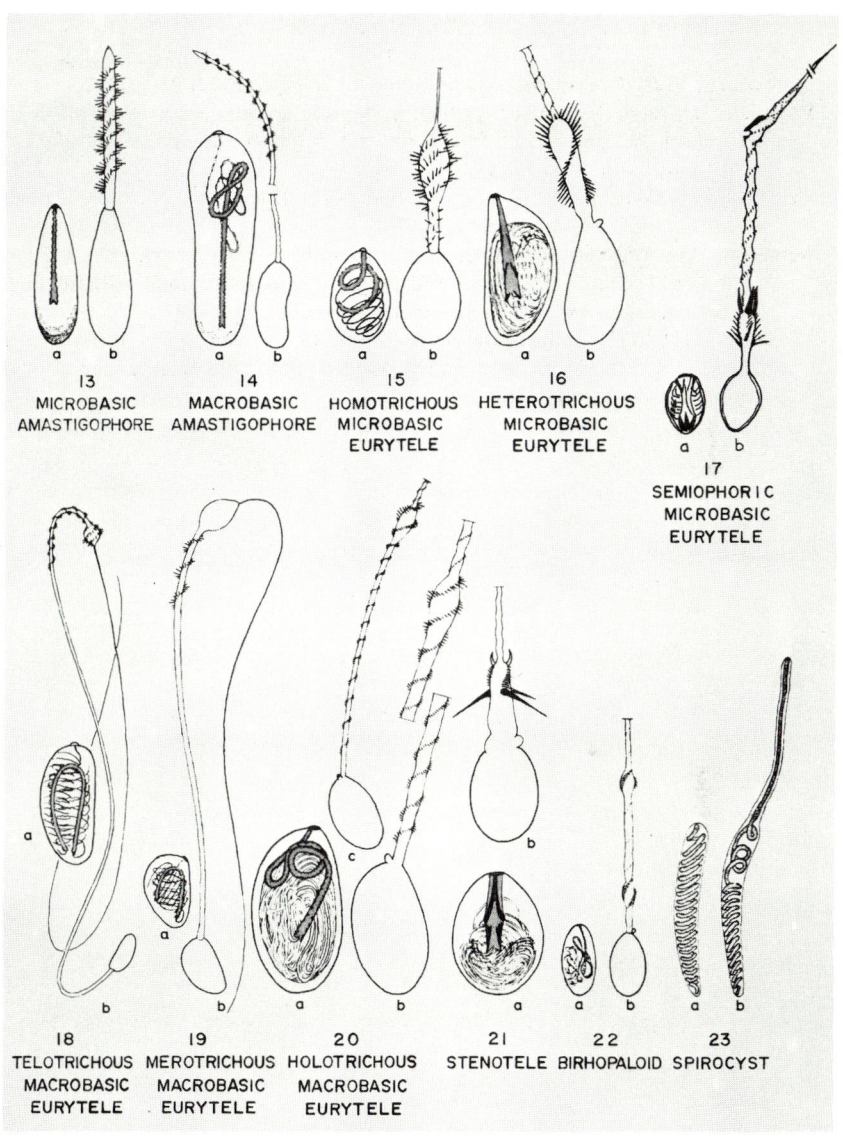

FIGS. 13-23. Coelenterate nematocysts.

Lenhoff, H. M., and E. S. Kline. 1958. The high imino acid content of the capsule from hydra nematocysts. *Anatomical Record* **130**: 425.

Lenhoff, H. M., E. S. Kline, and R. Hurley. 1957. A hydroxyproline-rich collagen-like protein of *Hydra* nematocysts. *Biochimica et biophysica acta* **26**: 204–205.

Mackie, G. O., and G. V. Mackie. 1963. Systematic and biological notes on living hydromedusae from Puget Sound. *Bulletin. National Museum of Canada,* no. 199, Contributions to zoology.

Weill, R. 1934. Contribution à l'étude des cnidaires et de leurs nématocystes. *Travaux de la Station Zoologique Wimereux* **10**: 11.

Werner, B. 1965. Die Nesselkapseln der Cnidaria, mit besonderer Berücksichtigung der, Hydroida. I. Klassifikation und Bedeutung für die Systematik und Evolution. *Helgoländer wissenschaftliche Meeresuntersuchungen* **12**:1–39.

Yanagita, T. M. 1959. Physiological mechanism of nematocyst responses in sea anemone. I. Effects of trypsin and thioglycolate upon the isolated nematocysts. *Japanese J. Zoology* **12**: 361–375.

DENNIS A. POWERS
University of Kansas, Lawrence

CHAPTER 16

Glucose-6-Phosphate Dehydrogenase and 6-Phosphogluconate Dehydrogenase Activities in Coelenterates

The demonstration of glucose-6-phosphate dehydrogenase (G6PDH) and 6-phosphogluconate dehydrogenase (6PGDH) activity in a tissue is considered strong evidence for the existence in that tissue of the hexose monophosphate shunt pathway (Hollman 1964). Using seven species of hydras, Rutherford and Lenhoff (1969) found considerable G6PDH activity, although they could not detect 6PGDH activity. The substrate 6-phosphogluconate, however, could be degraded by a phosphatase in a homogenate from *Hydra littoralis*, giving inorganic phosphate and gluconic acid (Rutherford and Lenhoff 1969).

I surveyed a wide variety of coelenterates at the Hawaii Institute of Marine Biology in order to see if the distribution of the two enzyme activities observed in hydras were specific for the Hydridae, or whether similar activities could be found in other coelenterates.

MATERIALS AND METHODS

All of the marine animals were collected off Coconut Island in Kaneohe Bay. Most of them were used as soon as collected. *Aiptasia*, however, was maintained in the laboratory for 3 or 4 weeks before use, was fed daily on nauplii of the brine shrimp *Artemia salina* and was kept in seawater changed twice daily. Specimens of the aposymbiotic (albino) *Aiptasia* were prepared by keeping the animals in the dark at 30° C for approximately 10 days (V. Buchsbaum Pearse and C. Cook, unpublished). *Boloceroides* was also maintained in the laboratory, but was kept in artificial seawater ("Instant Ocean," Aquarium Systems, Inc., Wickliffe, Ohio) prepared without the trace element supplement). The specimens of the aposymbiotic *Boloceroides* used had spontaneously lost their zooxanthellae while in the laboratory. The hydras (kindly supplied by Dr. L.

An earlier version of this chapter appeared as "Glucose-6-phosphate dehydrogenase and 6-phosphogluconate dehydrogenase activities in coelenterates" by Dennis A. Powers, Howard M. Lenhoff, and Charles A. Leone, in *Comparative Biochemistry and Physiology* 27: 139–144.

Davis) were laboratory-grown in "M" culture solution (Muscatine and Lenhoff 1965) by the methods of Loomis and Lenhoff (1956). The *Cordylophora* (kindly supplied by Dr. C. R. Wyttenbach) was grown in CCS-5 culture solution (Fulton 1960).

The extraction procedure to obtain soluble enzymes from the organisms was specially devised to remove the endosymbiotic zooxanthellae so common in the gastrodermises of many marine coelenterates. The tissue (or group of animals) was gently homogenized in a cold, loose-fitting, glass tissue-grinder. The resultant homogenate was centrifuged for 30 seconds in a clinical centrifuge to rid it of algae and any heavy unbroken material; the supernatant suspension was then centrifuged at approximately 18,000 X g in an RC-2 refrigerated centrifuge. The clear supernatant was assayed for G6PDH and 6PGDH activities.

The G6PDH and 6PGDH activities were measured fluorometrically by the methods of Lowry (1958a, b) and of Lowry, Roberts, and Kapphahn (1956). All readings were corrected, when necessary, for endogenous reduced nicotinamide adenine dinucleotide phosphate (NADPH) formation or disappearance, for background fluorescence, and for quenching of NADPH fluorescence by the coelenterate extracts.

The fluorometer (Farrand, model A) was fitted with a primary filter no. 7-37 and secondary filters nos. 3-73 and 4-70. The NADP, NADPH, tris-(hydroxymethyl)-aminomethane, 2-amino-2 methyl, 1,3-propanediol, glucose-6-phosphate, 6-phosphogluconate, crystalline G6PDH, and crystalline 6PGDH were purchased from Sigma Chemical Company, St. Louis, Mo. Protein was determined by the colorimetric method of Lowry et al. (1951) against a standard of bovine serum albumin (Armour and Company, Kankakee, Ill.).

Specific activities of the enzymes were expressed as micromoles of substrate converted per hour per microgram protein nitrogen (μmoles/hr per μg protein N).

The NADP and NADPH were standardized as described by Lowry, Roberts, and Kapphahn (1956) in a Beckman DB spectrophotometer. The glucose-6-phosphate and 6-phosphogluconate were standardized enzymatically by employing crystalline G6PDH and 6PGDH, respectively, and by following the stoichiometry of NADPH formation (Lowry, Roberts, and Kapphahn 1956).

The rate of disappearance of 6-phosphogluconate was measured by incubating 6 μl of 0.1 M 6-phosphogluconate, 10μl of extract, and 1 ml of tris-(hydroxymethyl)-amino-methane buffer (0.1 M, pH 8.2) for various periods from 0 to 120 minutes. The reactions were stopped by immersing the tubes in boiling water for 10 minutes. The amount of 6-phosphogluconate remaining unmetabolized by the homogenate was determined by measuring the amount of NADPH formed when NADP and crystalline 6PGDH were added to the cooled tubes.

RESULTS

Specific activities of G6PDH and 6PGDH were determined for 21 cnidarians and one ctenophore (Table 1). The results show that all extracts assayed had measurable G6PDH activity. The hydroids (except *Pennaria tiarella*) were high in G6PDH activity. The other coelenterates had generally lower and varying amounts of G6PDH activity.

The activity of 6PGDH was generally low in comparison to G6PDH activity. The most noticeable finding was the absence of detectable 6PGDH in all the extracts prepared from specimens of the order Hydroida. In addition, extracts of two of the three scyphozoans tested had no detectable 6PGDH activity; the third had a 6PGDH level of less than one-fifth of its G6PDH activity. The extracts of the other hydrozoan tested, the siponophore *Physalia utriculus*, also had a low level of 6PGDH activity and could metabolize 6PG as well in a system containing no added pyridine nucleotides (Table 2).

DISCUSSION

These data suggest that the findings of Rutherford and Lenhoff (1969) showing that hydra extracts lack 6PGDH activity might apply to all hydroids, whether they be marine or freshwater. Assays for 6PGDH in extracts of still more hydroids should prove interesting. Whether or not the absence of relatively low amounts of 6PGDH activity in the scyphozoans implies a close evolutionary relationship to the hydrozoans remains speculative.

Although G6PDH and 6PGDH activities are used as criteria for relative activity of the pentose shunt, an absence of either enzyme does not indicate an absence of the shunt (Hollman 1964). There are several alternative routes for glucose metabolites to enter the hexose monophosphate shunt and there are other ways in which tissues can utilize 6PG. Rutherford and Lenhoff (1969) demonstrated that preparations from *Hydra littoralis* can degrade 6PG to gluconic acid and inorganic phosphate. In the present study, I was able to demonstrate the disappearance of 6PG by preparations obtained from two hydras, a marine colonial hydroid, and three scyphozoans (Table 2). It would seem reasonable to suppose that such organisms could catabolize 6PG in a manner similar to *H. littoralis*.

If we assume that extracts of the albino *Chlorohydra viridissima* might form gluconic acid from 6PG, it is of interest that *C. viridissima* extracts could not metabolize gluconic acid in any of the NAD- or NADP-dependent, or ATP-dependent routes tested. Rutherford and Lenhoff (1969) demonstrated that extracts of *Hydra littoralis* do not metabolize 6PG via the Entner-Duodoroff pathway (1952).

TABLE 1 G6PDH and 6PGDH activities in 21 coelenterates and in one ctenophore

Experiment	Organisms[a]	Mean Values of Specific Activities × 10⁻⁴ µmoles/hr per µg Protein N		Mean Number of Individuals or Colonies per Trial	Number of Runs
		G6PDH	6PGDH		
	Class Hydrozoa				
	Order Hydroida				
1a	*Chlorohydra viridissima* (albino)	171.2	0	20	10
1b	*Chlorohydra viridissima*	137.5	0	20	5
2	*Hydra pseudo-oligactis*	178.3	0	20	7
3	*Pennaria tiarella*	0.4	0	10	3
4	*Obelia* sp.	195.8	0	30	2
5	*Plumularia* sp.	230.8	0	25	3
6	*Cordylophora lacustris*	91.3	0	4	2
	Order Siphonophora				
7	*Physalia utriculus*	92.4	18.0	1	2
	Class Scyphozoa				
8	*Carybdea* sp.	116.2	18.7	1	2
9	*Cassiopeia* sp.	31.7	0[b]	1	2

	Order Actinaria			
11	*Macranthia cookei*	6.4	19.4	1	3
12	*Anthopleura* sp.	6.3	24.5	4	2
13	*Anthopleura nigrescians*	2.0	1.8	5	2
14a	*Aiptasia* sp. (albino)	18.1	18.2	2	10
14b	*Aiptasia* sp.	18.0	18.2	2	10
15a	*Boloceroides* sp. (albino)	90.9	21.9	2	4
15b	*Boloceroides* sp.	69.1	28.6	2	4
16	*Boloceroides lilae*	1.5	15.7	1	2
	Order Zoanthidea			
17	*Zooanthus* sp.	31.2	23.4	4	2
18	*Palythoa* sp.	24.6	24.6	6	3
	Order Madreporaria			
19	*Fungia scutaria*	4.1	3.8	1	3
20	*Pocillopora damicornis*	33.3	22.9	1	3
21	*Dendrophyllia manni* (*Tubastrea aurea*)	4.6	3.7	1	4
	Class Ctenophora			
22	*Cydippida*	180.0	18.0	1	3

[a] Classification of organisms according to the taxonomy of Hyman (1940).
[b] Possibly a trace of 6PGDH activity which may be hidden by endogenous NADPH production.

TABLE 2   Rate of disappearance of 6PG in the absence of pyridine nucleotides

Experiment	Species	Rate of Disappearance of 6PG $\times$ 10^{-4} μmoles/Hour per μg Protein N
1a	Chlorohydra viridissima (albino)	60.7
2	Hydra pseudo-oligactis	83.3
4	Obelia sp.	26.3
8	Carybdea sp.	5.3
9	Cassiopeia sp.	5.0
0	Mestigius sp.	44.7

NOTE: Extracts of aposymbiotic Chlorohydra viridissima had no activity for gluconoreductase, NAD- or NADP-dependent glucose oxidase, δ-gluconolactone reductase, NAD- or NADP-dependent gluconic oxidase, nor for ATP-dependent gluconokinase.

The possibility that some endosymbiotic algae were disrupted and leaked G6PDH or 6PGDH activity into the animal extracts does not seem likely. There was no 6PGDH activity in extracts prepared from symbiotic Chlorohydra viridissima (Table 1, experiment 1b). In addition, extracts prepared from the aposymbiotic Aiptasia sp. (Table 1, experiment 14a) and Boloceroides sp. (Table 1, experiment 15a) showed no significant differences in specific activities from extracts prepared from symbiotic specimens of these two species (Table 1, experiments 14b and 15b). If the specific activities of the enzymes in actinarians and the enzymes of their endosymbiotic zooxanthellae were not the same, then leakage of the enzymes from the algae should have significantly altered the specific activities observed; such was not the case.

My results show that 6PGDH is absent, not only in the Hydridae (Rutherford and Lenhoff 1969), but in calyptoblastic and gymnoblastic hydroids and some scyphozoans as well. Because the pentose shunt pathway in some cnidarians appears to differ from that of most organisms studied, further investigation of carbohydrate metabolism in the Hydrozoa and Scyphozoa should prove interesting.

SUMMARY

1. Tissue extracts from 21 cnidarian species and one ctenophore were assayed for glucose-6-phosphate dehydrogenase and 6-phosphogluconate dehydrogenase activities.

2. All extracts had glucose-6-phosphate dehydrogenase activity, the hydroids being most active.

3. There was no measurable 6-phosphogluconate dehydrogenase activity in any of the hydroid extracts tested nor in the extracts obtained from two of three species of scyphozoans.

4. The substrate 6-phosphogluconate was shown to be metabolized by extracts from three hydroids and three scyphozoans tested, presumably by a phosphatase.

5. Extracts of aposymbiotic *Chlorohydra viridissima* had no activity for gluconoreductase, NAD- or NADP-dependent glucose oxidase, δ-gluconolactone reductase, NAD- or NADP-dependent gluconic oxidase, nor for ATP-dependent gluconokinase.

LITERATURE CITED

Entner, N., and M. Doudoroff. 1952. Glucose and gluconic acid oxidation of *Pseudomonas sacharophila*. *J. Biological Chemistry* **196**(2): 853–862.

Fulton, C. 1960. Culture of colonial hydroid under controlled conditions. *Science* **132**: 473–474.

Hollman, S. 1964. *Non-glycolytic pathways of metabolism of glucose*, pp. 57–60. New York: Academic Press.

Hyman, L. H. 1940. *The invertebrates: Protozoa through Ctenphora*, vol. 1, pp. 356–641. New York: McGraw-Hill.

Loomis, W. F., and H. M. Lenhoff. 1956. Growth and sexual differentiation of hydra in mass culture. *J. Experimental Zoology* **132**: 555–574.

Lowry, O. H. 1958a. The quantitative histochemistry of brain. V. Enzymes of glucose metabolism. *J. Biological Chemistry* **232**: 979–993.

———. 1958b. The quantitative analysis of single nerve cell bodies. In *Ultrastructure and cellular chemistry of neural tissue*, H. Waelsch, ed., pp. 69–76. New York: Hoeber-Harper.

Lowry, O. H., N. R. Roberts, and J. L. Kapphahn. 1956. The fluorometric measurement of pyridine nucleotides. *J. Biological Chemistry* **224**(2): 1047–1064.

Lowry, O. H., N. J. Rosenbrough, A. L. Farr, and R. J. Randall. 1951. Protein measurement with the Folin phenol reagent. *J. Biological Chemistry* **193**: 265–275.

Muscatine, L., and H. M. Lenhoff. 1965. Symbiosis of hydra and algae. I. Effects of some environmental cations on growth of symbiotic and aposymbiotic hydra. *Biological Bulletin, Woods Hole* **128**: 415–424.

Rutherford, C., and H. M. Lenhoff. 1969. Enzymes of glucose catabolism in hydra. I. Relative activites of enzymes and absence of 6-phosphogluconate dehydrogenase. *Archives of Biochemistry and Biophysics* **133**: 119–127.

PART 3

ENDOSYMBIOSIS WITH ALGAE

LEONARD MUSCATINE
University of California at Los Angeles

CHAPTER 17

Endosymbiosis of Algae and Coelenterates

Endosymbiosis with algae is widespread among invertebrates. About 150 genera of marine and freshwater invertebrates, representing eight phyla, possess algal symbionts. The algae are generally referred to as either "zoochlorellae" if they are green and inhabit freshwater invertebrates, or "zooxanthellae" if they are brown and inhabit marine invertebrates. More precise taxonomic connotations are now known and will be discussed below (see Smith, Muscatine, and Lewis 1969). Amongst coelenterates, only one genus (*Chlorohydra*) contains zoochlorellae (*Chlorella* sp.); all others possess zooxanthellae.

The main purpose of this review is to discuss some unsolved problems of coelenterate symbiosis, especially symbiosis with zooxanthellae. Descriptive aspects will be only briefly considered, since there is a recent general review of "endozoic algae" by McLaughlin and Zahl (1966), and a review of the biology of coral reefs by Yonge (1963).

ASSOCIATION WITH ZOOCHLORELLAE

Morphological Aspects

Amongst coelenterates, zoochlorellae exist only in the freshwater hydras, where they are confined exclusively to the gastrodermal cells. The algae (Chlorococcales) are green, unicellular, 6–12 microns in diameter, ellipsoidal, and possess a cup-shaped chloroplast, a pyrenoid, and a thin cell wall. Typically, within each gastrodermal cell there are 15–25 algae. The specificity of their location within the gastrodermal cell was demonstrated by Haynes and Burnett (1963). They isolated pieces of hydra gastrodermis and observed that some cells differentiated into epidermal cells which rejected their symbiotic algae; those which remained as gastrodermal cells retained the algae.

The algae reproduce within the host by formation of autospores (Park et al. 1967; Oschman 1967). Algal daughter cells are passed on to hydra offspring via buds during asexual reproduction. Hydra eggs often contain algae (Brien and Reniers-Decoen 1950). Eggs form from epidermal gonads; the epidermis itself is free of algae. It is not known precisely how the algae "infect" a hydra cell, or how the numbers of algae in a host cell are regulated. Although the algae continue active photosynthesis for at least 24 hours after isolation from the host, attempts to culture them have failed (Muscatine 1965). Successful culture of zoochlorellae from *Paramecium bursaria* and *Spongilla* sp. (see Muscatine, Karakashian, and Karakashian 1967) raises the hope that the technical difficulties of culturing the hydra algae will be eliminated. Some strains of *Chlorohydra* are easily rid of their symbionts by incubation for 8 days with 0.5 percent glycerol (Whitney 1907; Muscatine and Lenhoff 1965*b*). The mechanism by which glycerol acts to release the algae is not known. Not all *Chlorohydra* strains are susceptible to this treatment.

Physiology

As early as 1882 Geddes proposed that the algal symbionts of hydra might supply oxygen to their hosts and use the carbon dioxide given off by the respiratory activities of the animal. He also reasoned that the algae might manufacture carbohydrates and pass on excess soluble materials to the animal host in much the same way as a chloroplast interacts with colorless cells of a plant leaf. Recent technical advances have permitted critical testing of Geddes' hypotheses.

One of the most important of these advances was the development of culture techniques for hydras by Loomis (1953, 1954). These techniques enabled investigators to maintain an unlimited number of experimental and aposymbiotic animals of known developmental, genetic, and nutritional history in the laboratory under controlled environmental conditions. The overall results of the ensuing quantitative experimental studies on symbiosis may be summarized briefly. When fed daily, both green and aposymbiotic hydra (Carolina strain) grow at nearly identical clonal rates, although the aposymbiotic hydra are somewhat smaller than are the green hydra. Thus, the algae are not essential for growth and survival of the hydra as long as the host is fed daily. However, during periods of starvation or when food is given intermittently, the clonal growth rate and survival of the green hydra is significantly greater than that of their aposymbiotic counterparts (Muscatine and Lenhoff 1965*a, b*). In fact, when green hydra are starved, the contribution of the algae to hydra growth efficiency (percent of energy consumed as food that is converted to new protoplasm) (Slobodkin 1962) actually increases (Stiven 1965). Because the aposymbiotic hydra do not need algae when food is abundant, it can be concluded that the algae do not augment growth and survival by producing oxygen or removing carbon dioxide. Rather, the algae probably contribute organic materials to enhance the growth and survival of the host.

The photosynthetic products excreted by the algae associated with hydra were identified by using isolated zoochlorellae. Such isolated algae, as well as cultures of the endosymbiotic algae from *Paramecium bursaria* and *Spongilla*, excreted up to 80 percent of the carbon fixed during photosynthesis, mostly as maltose (Muscatine 1965; Muscatine, Karakashian, and Karakashian 1967). In contrast, free-living *Chlorella* excrete only a small proportion of their fixed carbon, mainly as glycollic acid. Using $^{14}CO_2$, it has been shown that photosynthetically fixed carbon is rapidly translocated in the intact hydra from the algae in the gastrodermis to the algae-free epidermis (Muscatine and Lenhoff 1963; Eisenstadt, this volume). The host tissue is rich in maltase (Muscatine 1965; Muscatine, Karakashian, and Karakashian 1967) and studies with radioactive carbon have demonstrated that ^{14}C received from the algae is incorporated into a wide range of substrates in the animal, including glycogen and pentose sugars (Muscatine and Lenhoff 1963; Roffman and Lenhoff 1969).

Although it is now firmly established that the algae can supply fixed carbon to the animal, many problems remain to be worked out. For example, why are the algae confined to the gastrodermal cells? What controls cell division of the algae so that they do not overgrow their host? Does the host influence photosynthesis and carbohydrate excretion by the algae? Why can some strains of *Chlorella* enter into symbiotic associations and other strains cannot? To what extent is transfer of nutrients (translocation) carried out from animal cells to the algae?

ASSOCIATIONS WITH ZOOXANTHELLAE

Morphological Aspects

Zooxanthellae in coelenterates are now acknowledged to be a stage in the life cycle of a species of dinoflagellate (McLaughlin and Zahl 1966; Freudenthal 1962; Jeffrey and Haxo 1968). Zooxanthellae are prevalent in marine coelenterates that inhabit a wide range of latitudes and depths, and are of profound ecological importance to reef corals with which they are invariably associated.

The symbionts are spherical, 8-12 microns in diameter, and most occur singly or in groups of two or three within the gastrodermal cells of adult hosts. Exceptions to this intracellular location have been catalogued by Droop (1963). Goreau (1961b) stated that zooxanthellae in reef corals are housed within special "carrier cells." Kawaguti (1964), on the other hand, concluded from electron microscopical studies on the coral *Oulastrea* that the algae are intercellular. The cytochemistry and ultrastructure of the symbionts in the anemone *Anemonia sulcata* are given by Taylor (1968).

Zooxanthellae from a variety of coelenterates have been maintained in culture. Those from the jellyfish *Cassiopeia* give rise to a variety of cell types including a motile gymnodinioid (Freudenthal 1962). The cultured cells differ in morphology from symbiotic forms. For example, electron micrographs of zooxanthellae cultured from *Cassiopeia* show a complex cell envelope of unusual thickness with intramembranal spaces; by contrast, native symbiotic cells do not show a stout wall (McLaughlin and Zahl 1965). Zooxanthellae reproduce relatively slowly by fission within the host. Little is known of their biology as free-living dinoflagellates, nor are there critical observations on the mode by which they infect host tissue. For recent reviews, see those of McLaughlin and Zahl (1966) and Yonge (1963).

Vertical Distribution of the Association in the Ocean

The majority of the coelenterate-zooxanthellae associations, especially hermatypic corals, occur in the upper 30 meters of the ocean. Although it is generally agreed that light is the primary environmental factor limiting vertical distribution, some important exceptions demonstrate the need for more critical experimental data. For example, anemones with zooxanthellae have been collected in Antarctica from 207 fathoms by Stephenson (1910), and both anemones and corals with zooxanthellae have been dredged from 100–116 fathoms of Key Largo, Florida by McLaughlin and Zahl (1959). In both cases, the hosts were surely below the compensation point for photosynthesis. The observations raise questions concerning the heterotrophic capabilities of zooxanthellae. Attempts to grow zooxanthellae in darkness have thus far failed, but cultures of zooxanthellae or algae in hosts maintained in darkness remain viable for 50–75 days (McLaughlin and Zahl 1959; Muscatine 1961*a*) and then exhibit active photosynthesis when returned to the light. It would be desirable to study the photosynthetic capacity of zooxanthellae, *in situ,* as a function of depth.

Production and Consumption of Oxygen and Carbon Dioxide

Numerous investigators have shown that, in daylight or with optimum illumination, algae-animal associations produce more oxygen than they consume (Sargent and Austin 1949; Odum and Odum 1955; Kohn and Helfrich 1957; Kanwisher and Wainwright 1967; Roffman 1968). The oxygen produced by the algae is presumably available to the host, but, as pointed out by Yonge (1963) and Droop (1963), its quantitative importance is uncertain since few associations are found in situations where oxygen content is abnormally low. Further, coelenterates deprived of their zooxanthellae have not yet been reported to suffer from effects attributable to lack of oxygen. The extent to which an aposymbiotic coelenterate can withstand anaerobiosis compared to one with symbiotic algae clearly ought to be investigated experimentally.

Some coelenterates with algae (as well as associations of algae with *Paramecium*) display phototaxis not exhibited by their aposymbiotic controls (Zahl and McLaughlin 1959). Thus, some authors suggest that the animal senses a gradient of oxygen rather than one of light intensity (e.g., see Droop 1963); experimental verification of this assumption, however, is lacking. The action spectrum for such a phototaxis might indicate the extent to which the algae influence this behavior.

It is usually assumed that symbiotic algae can use host respiratory carbon dioxide for photosynthesis, especially because the host contains a substantial bicarbonate pool (Goreau 1961a). In addition, the algae may potentially draw on a virtually unlimited supply of carbon dioxide in the ambient oceanic environment. The extent to which the photosynthetic requirements of the algae are satisfied by exogenous vs. endogenous sources of carbon dioxide is not yet clear. Experiments in which the host is given food labeled with ^{14}C and then its algae are assayed for labeled photosynthetic products may contribute some information as to whether the algae utilize endogenous carbon dioxide. In such experiments, it is important to determine the extent to which other ^{14}C-labeled substances are transferred from the host to the algae. Utilization of endogenous CO_2 by algae in corals may affect the rates of calcification in reef corals.

According to Yonge and Nicholls 1931 and Kawaguti 1953, the host may benefit by having its nitrogenous and phosphatic wastes removed by the algae. Although it has been demonstrated that ammonia and phosphate are removed from solution by corals, any advantage to the host still awaits experimental verification.

Role of the Algae in the Nutrition of the Host

The extent to which zooxanthellae supply nutrients to the host has often been disputed. From observations on sea anemones and reef-building corals, Yonge and Nicholls (1931) suggested that zooxanthellae do not contribute substantially to organic productivity of their host, noting that (1) there is adequate plankton to support the needs of the host so that zooxanthellae are not an essential food source; (2) corals and anemones feed exclusively on zooplankton, rejecting plant material; (3) corals have enzymes specialized for a carnivorous diet; (4) starved corals eject their zooxanthellae; and (5) corals can live in darkness (Yonge 1957).

More recent investigations now suggest that ejection of zooxanthellae by starving corals may result from several different factors, including decrease in space as the host's tissues shrink, change in salinity (Goreau 1964), or elevated temperature (Cook, unpublished).

Although corals may appear to thrive in darkness or without algae, the critical assessment of their well-being must come from quantitative measurements of growth of their tissues. Muscatine (1961a, b) observed that starved anemones with algae lose weight less rapidly than starved aposymbiotic anemones, in either light or darkness, despite the fact that aposymbiotic anemones appeared perfectly healthy and those with algae ejected them as the tissues regressed. Because of their less rapid weight loss, the anemones with algae have an obvious selective advantage over the aposymbiotic algae. Such observations illustrate the need for an objective, quantitative evaluation of the well-being of the host in assessing the role of algae in host nutrition. Recent quantitative studies show that zooxanthellae in some associations may well be of major importance to host nutrition. Two modes of nutritional augmentation have been proposed: (1) digestion of zooxanthellae, and (2) utilization of products released from living zooxanthellae *in situ*.

Digestion of Zooxanthellae

Digestion of algae is usually inferred from observations of fragments of algae within host cells. For example, Kawaguti (1965) showed by electron microscopy that zooxanthellae from the coral *Oulastrea* have a thin plasma membrane covered with an accumulation of double membrane fragments. He concluded that these fragments were available for assimilation by the host. Critical evidence that the host gains nutritional benefit from the algal fragments is lacking.

McLaughlin et al. (1963) reported that axenic cultures of zooxanthellae from *Cassiopeia* excrete insoluble extracellular mucoid substances which yield glucose on hydrolysis, and they speculated that this mucoid material may be a potential source of carbon for the host. In this case, it is not known whether the mucoid substance is produced by the algae while they are within their host, or whether the cells only produce the mucoid material when they are cultured outside of the host. Production of insoluble substances after removal from the host has been observed in the case of lichen algae (Richardson, Hill, and Smith, unpublished).

Utilization of Products of Photosynthesis Released by Zooxanthellae

There is now abundant evidence that coelenterates acquire soluble organic carbon compounds released from zooxanthellae during photosynthesis. In a variety of experiments in which $^{14}CO_2$ is given to symbiotic coelenterates in the light, a substantial proportion of the ^{14}C is invariably found in the host animal's tissues.

One of the earliest demonstrations of this transfer was made by Muscatine and Hand (1958) using the sea anemone *Anthopleura elegantissima*. Anemones with zooxanthellae were placed in seawater containing $NaH^{14}CO_3$. In the light, the algae incorporated some of the labeled carbon into organic molecules whose initial location and subsequent fate were determined by radioautography. Anemones in labeled seawater, but in the dark, served as controls for heterotrophic fixation of labeled carbon. Translocation of label from algae to animal tissue was verified in radioautographs of sections of animals that were incubated in $^{14}CO_2$ for 1–5 weeks. No attempt was made in that study to determine the nature or quantity of the translocated materials. Goreau and Goreau (1960) performed similar experiments with the corals *Manicina areolata* and *Montastrea annularis*. They cited evidence for translocation after 50 hours' incubation but stated that "it appears probable that the amount transferred from the algae to the host can at best satisfy only a very small part of the coral's total nutritional requirement." Although autoradiography provides direct evidence for translocation, it is at best only a semiquantitative technique. Errors are introduced by the conventional processes of preparing sections because that material which is soluble in solvents used (alcohol, water, toluene, etc.) is removed, leaving only insoluble material. Since the relative proportions of radioactive material in soluble and insoluble material may vary in each preparation, the extent of translocation may easily be misinterpreted. In fact, if translocation in corals involved exclusively soluble materials, this process would be impossible to detect by conventional autoradiography.

Taylor (1968) observed that algae in *Anemonia sulcata* have a complex periplast consisting of four to five membranes displaying folds and vesicles. Since autoradiography of ^{14}C-labeled sections of *Anemonia* did not yield conclusive evidence for translocation, Taylor speculated that the periplast acted as a barrier to passage of organic molecules. However, in more recent experiments, Taylor (1969) did detect translocation in *Anemonia*. Perhaps the complex periplast with its increased surface area might well be interpreted as an adaptation for absorption of materials from the host.

Quantitative aspects of translocation in *Anthopleura* have been investigated by R. K. Trench (1969). After incubating excised tentacles in seawater containing $NaH^{14}CO_3$ for 10 hours, he demonstrated that the algae-free epidermal tissues accumulate about 30 percent of the total labeled carbon fixed by the tentacle. Fractionation of the epidermis revealed that the labeled material appears in both the alcohol-soluble and alcohol-insoluble fractions of the host tissue.

Direct evidence that zooxanthellae from corals and anemones can release photosynthetically produced organic compounds comes from studies using algae immediately after they have been removed from the host (Muscatine 1967; Trench 1968). Algae excrete up to 60 percent of the total carbon fixed in photosynthesis, mostly as glycerol, accompanied by small amounts of other compounds. Trench (1968) further showed that the nature of the substances excreted and the extent of excretion changes soon after the algae are isolated. Von Holt and von Holt (1968*a, b*) described both the secretion of a wide range of soluble organic materials by algae from *Zoanthus,* after the algae had been isolated and incubated *in vitro* for several hours, as well as *in vivo* translocation in anemones, zoanthids, and a coral.

Some host tissues stimulate excretion of glycerol by zooxanthellae

Zooxanthellae isolated from the reef coral *Pocillopora damicornis* and incubated *in vitro* with $Na_2{}^{14}CO_3$ released less than 5 percent of the total carbon fixed into the medium after 2–4 hours incubation. However, when a homogenate of host tissue was added to the incubation mixture, 30–50 percent of the total carbon fixed was rapidly released to the medium, principally as glycerol. The release of glycerol was proportional to the amount of host homogenate added, and boiling the homogenate abolished its ability to stimulate the algae to release glycerol. Algae from a variety of other reef corals, a xeniid soft coral, and the giant clam *Tridacna* also released glycerol in the presence of host tissue (Muscatine 1967). Trench (1968) obtained similar results with algae from a subtidal zoanthid (*Anthopleura*), and a tropical scyphozoan (*Cassiopeia*). He also demonstrated that a homogenate of *Anthopleura* tissue stimulated release of glycerol from its algae *in vitro*, but that homogenate of aposymbiotic *Anthopleura* did not. However, when the latter was reinfected with algae, the homogenate displayed the stimulatory activity.

Utilization of translocated carbon by the reef coral *Pocillopora damicornis* has been investigated by Muscatine and Cernichiari (1969). After 24 hours' incubation on the reef under natural diurnal conditions, translocation *in vivo* ranged from 36 to 50 percent of the total $^{14}CO_2$ fixed. The translocated ^{14}C appeared in the host lipid, protein, skeletal carbonate, and skeletal organic matrix. Thus, in at least two zooxanthellae associations (*Anthopleura* and *Pocillopora*), the nature and extent of translocation has been experimentally elucidated.

It is interesting to note that Hellebust (1965) studied excretion of photoassimilated radiocarbon in 22 species of free-living marine phytoplankton. Most retained their photosynthate and liberated only glycolic acid. Glycerol was excreted in very small amounts by free-living species of Bacillariophyceae, Chlorophyceae, and Cyanophyceae. Free-living dinoflagellates, however, did not excrete glycerol. Excretion by symbiotic algae of carbohydrates which are either absent in free-living relatives or present in only small amounts is a widespread phenomenon among different symbionts (Smith, Muscatine, and Lewis 1969). The general phenomenon of release and translocation of carbohydrate is now known to occur in symbiotic associations with fungal pathogens and angiospermous parasites and in lichens (Smith, Muscatine, and Lewis 1969).

The importance of translocated carbohydrate from symbiotic algae to their coelenterate hosts seem primarily nutritional. A "built-in" supply of reduced organic carbon at presumably low energy-cost would be advantageous to the host where food is limited, as is the case in tropical ocean waters. Only supplementary zooplankton might be needed to provide sufficient inorganic nutrients such as nitrogen and phosphorus for host maintenance metabolism and synthetic processes. Quantitative feeding experiments with green hydra support this view (Muscatine and

Lenhoff 1965b), and analogous experiments should be carried out on zooxanthellae associations.

Although coelenterates may obtain soluble organic materials from their plant symbionts, they may still feed as conventional carnivores. On the other hand, much remains to be learned about the apparent lack of conventional carnivorous feeding behavior of many coelenterates that harbor zooxanthellae. For example, many species of the orders Alcyonacea (*Xeniidae*), Gorgonacea, and Zoanthidea have not yet been observed to feed as conventional carnivores (Gohar 1940, 1948; Wainwright 1967; Reimer, this volume). In fact, many of these hosts expand their polyps in the daytime when light is plentiful and plankton relatively sparse, and contract them at night, when plankton is relatively plentiful but light unavailable. Wainwright (1967) suggests that this behavior indicates nutritional dependence on zooxanthellae during the daytime and less dependence on zooplankton food sources. Questions of nitrogen economy in these associations still remain unanswered.

The advantages of the association to the algae may also be nutritional. Since the algal symbionts of most coelenterates are intracellular, the nutrients which they require must move through, or emanate from, the animal cells. In nutrient-poor tropical waters, access to nutrients derived from digestion of food by the host would be of great advantage to algae. Because such a source of nutrients would not be immediately available to free-living algae, the density of symbiotic algae in their hosts is frequently greater than the density of free-living algae in the surrounding seawater.

On the other hand, the metabolic activities of the host may control the growth of the symbionts by regulating the supply of some essential algal nutrient or growth factor. Yonge (1963), for example, has suggested that the supply of nitrogen and phosphorus may limit the growth of zooxanthellae *in situ*; however, some experimental investigation would be required to test whether the host's carrier cells could tolerate the low level of nitrogen and phosphorus that would be required to inhibit the growth of the algae, especially since the algae can use a wide range of nitrogen and phosphorus compounds. Alternatively, the host may limit some factor(s) which are required for algal growth but which are not essential to the host cells. Finally, the host may secrete specific substances which control algal cell division *in situ*.

Role of Zooxanthellae in Coral-Reef Productivity

Because a reef community or isolated coral head produces more oxygen than it consumes in a 24-hour diurnal period (Sargent and Austin 1949, 1954; Odum and Odum 1955; Kohn and Helfrich 1957; Kanwisher and Wainwright 1967; Roffman 1968), presumably more organic material is produced than is consumed. Some studies (Sargent and Austin 1949,

1954; Odum and Odum 1955) suggest that the amount of plankton moving across reef platforms is inadequate to meet the requirements of the consumers, but that this deficit is offset by daily production of reduced organic carbon by algae, principally the symbiotic algae, on the reef. Early attempts to assign to zooxanthellae the role of primary producer were hindered by lack of evidence for the manner in which energy of primary producers was transferred to the corals and to the community and by inaccurate estimates on the reef and coral plant biomass. This lack of evidence resulted in an interpretation emphasizing the potential role of the filamentous algae in the skeleton as primary producers. Their biomass was thought to exceed that of the zooxanthellae by a factor of 16. Translocation phenomena now provide evidence to explain how reduced carbon from zooxanthellae may be acquired by corals, and recent work by Kanwisher and Wainwright (1967) and Halldal (1968) shows that photosynthesis by filamentous algae is negligible compared to that of zooxanthellae, inferring that the filamentous algae contribute little, if any, reduced carbon to the coral or to the community. The role of zooxanthellae in calcification of reef corals is discussed in this volume in Chapter 22.

Further studies on nutrition and metabolism of coelenterates with zooxanthellae should take into consideration the work of Stephens (1960) who demonstrated that some corals (and other invertebrates as well) accumulate soluble organic material from solution, and that the amount accumulated may satisfy part of the animal's maintenance needs. Some observations on accumulation of glycine from solution by zoanthids are presented by Reimer in this volume.

LITERATURE CITED

Brien, P., and M. Reniers-Decoen. 1950. Étude d'*Hydra viridis* (Linnaeus) (la blastogenese, la spermatogenese, l'ovogenese). *Annales, Société r. zoologique de Belgique* **81**: 33-110.

Droop, M. 1963. Algae and invertebrates in symbiosis. In *Symbiotic associations. Society for General Microbiology. Symposium,* no. 13, B. Mosse and P. Nutman, eds. Cambridge.

Freudenthal, H. 1962. *Symbiodinium* gen. nov. and *Symbiodinium microadriaticum* sp. nov., a zooxanthella: Taxonomy, life cycle, and morphology. *J. Protozoology* **9**: 45-57.

Geddes, P. 1882. The yellow cells of radiolarians and coelenterates. *Proceedings, Royal Society of Edinburgh* **11**: 377-396.

Gohar, H.A.F. 1940. Studies on the Xeniidae of the Red Sea. *Publications, Marine Biological Station, Ghardaqa* **2**: 25-118.

———. 1948. A description of some biological studies of a new alcyonarian species

Clavularia hamra Gohar. *Publications, Marine Biological Station, Ghardaqa* **6**: 1–33.
Goreau, T. F. 1961*a*. On the relation of calcification to primary production in reef building organisms. In *The biology of hydra and of some other coelenterates: 1961*, H. M. Lenhoff and W. F. Loomis, eds. Coral Gables: University of Miami Press.
———. 1961*b*. Problems of growth and calcium deposition in reef corals. *Endeavour* **20**: 32–39.
———. 1964. Mass expulsion of zooxanthellae from Jamaican reef communities after Hurricane Flora. *Science* **145**: 383–386.
Goreau, T.F., and N.I. Goreau. 1960. Distribution of labelled carbon in reef-building corals with and without zooxanthellae. *Science* **131**: 668–669.
Halldal, P. 1968. Photosynthetic capacities and photosynthetic action spectra of endozoic algae of the massive coral *Favia*. *Biological Bulletin, Woods Hole* **134**: 411–424.
Haynes, J., and A. Burnett. 1963. Dedifferentiation and redifferentiation of cells in *Hydra viridis*. *Science* **142**: 1481–1483.
Hellebust, J.A. 1965. Excretion of some organic compounds by marine phytoplankton. *Limnology and Oceanography* **10**: 192–206
Holt, C. von, and M. von Holt. 1968*a*. Transfer of photosynthetic products from zooxanthellae to coelenterate hosts. *Comparative Biochemistry and Physiology* **24**: 73–81.
———. 1968*b*. The secretion of organic compounds by zooxanthellae isolated from various types of *Zoanthus*. *Comparative Biochemistry and Physiology* **24**: 83–92.
Jeffrey, S.W., and F.T. Haxo. 1968. Photosynthetic pigments of symbiotic dinoflagellates (zooxanthellae) from corals and clams. *Biological Bulletin, Woods Hole* **135**: 149–165.
Kanwisher, J.W., and S.A. Wainwright. 1967. Oxygen balance in some reef corals. *Biological Bulletin, Woods Hole* **133**: 378–390.
Kawaguti, S. 1953. Ammonium metabolism of the reef corals. *Biological J. Okayama University* **1**: 171–176.
———. 1964. Zooxanthellae in the corals are intercellular symbionts. *Proceedings, Japan Academy of Science* **40**: 545–548.
———. 1965. An electron microscopic proof for a path of nutritive substances from zooxanthellae to reef coral tissue. *Proceedings, Japan Academy of Science* **40**: 832–835.
Kohn, A.J., and P. Helfrich. 1957. Primary organic productivity of a Hawaiian coral reef. *Limnology and Oceanography* **2**: 241–251.
Loomis, W.F. 1953. The cultivation of hydra under controlled conditions. *Science* **117**: 565–566.
———. 1954. Environmental factors controlling growth on hydra. *J. Experimental Zoology* **126**: 223–234.
McLaughlin, J., and P. Zahl. 1959. Axenic zooxanthellae from various invertebrate hosts. *Annals, New York Academy of Science* **77**: 55–72.
———. 1966. Endozoic algae. In *Symbiosis*, vol. 1, S. Mark Henry, ed., chapter 5. New York: Academic Press.

McLaughlin, J.A., P.A. Zahl, A. Nowak, and J. Marchisotto. 1963. Some constituents of zooxanthellae grown in axenic culture. In *Progress in protozoology.* Proceedings of the First International Congress on Protozoology, Prague, August 22 - 31, 1961, J. Ludvik, J. Lom, and J. Vávra, eds., pp. 204–205. New York: Academic Press.

Muscatine, L. 1961a. Some aspects of the relationship between a sea anemone and its symbiotic algae. Ph.D. dissertation, University of California at Berkeley.

──── . 1961b. Symbiosis in marine and fresh water coelenterates. In *The biology of hydra and of some other coelenterates: 1961*, H.M. Lenhoff and W.F. Loomis, eds., pp. 255–268. Coral Gables: University of Miami Press.

──── . 1965. Symbiosis of hydra and algae. III. Extracellular products of the algae. *Comparative Biochemistry and Physiology* **16**: 77–92.

──── . 1967. Glycerol excretion by symbiotic algae from corals and *Tridacna* and its control by the host. *Science* **156**: 516–519

Muscatine, L., and E. Cernichiari. 1969. Assimilation of photosynthetic products of zooxanthellae by a reef coral. *Biological Bulletin, Woods Hole* **137**: 506–523.

Muscatine, L., and C. Hand. 1958. Direct evidence for transfer of materials from symbiotic algae to the tissues of a coelenterate. *Proceedings, National Academy of Sciences* **44**: 1259–1263.

Muscatine, L., and H.M. Lenhoff. 1963. Symbiosis: On the role of algae symbiotic with hydra. *Science* **142**: 956–958.

──── 1965a. Symbiosis of hydra and algae. I. Effects of some environmental cations on growth of symbiotic and aposymbiotic hydra. *Biological Bulletin, Woods Hole* **128**: 415–424.

──── . 1965b. Symbiosis of hydra and algae. II. Effects of limited food and starvation on growth of symbiotic and aposymbiotic hydra. *Biological Bulletin, Woods Hole* **129**: 316–328.

Muscatine, L., S.J. Karakashian, and M.W. Karakashian. 1967. Soluble extracellular products of algae symbiotic with a ciliate, a sponge and a mutant hydra. *Comparative Biochemistry and Physiology* **20**: 1–12.

Odum, H.T., and E.P. Odum. 1955. Trophic structure and productivity of a windward coral reef community on Eniwetok atoll. *Ecological Monographs* **25**: 291–320.

Oschman, J.L. 1967. Structure and reproduction of the algal symbionts of *Hydra viridis. J. Phycology* **3**: 221–228.

Park, H., C.L. Greenblatt, C.F.T. Mattern, and C.R. Merril. 1967. Some relationships between *Chlorohydra,* its symbionts and some other chlorophyllous forms. *J. Experimental Zoology* **164**: 141–162.

Roffman, B. 1968. Patterns of oxygen exchange in some Pacific corals. *Comparative Biochemistry and Physiology* **27**: 405–418.

Roffman, B., and H. M. Lenhoff. 1969. Formation of polysaccharides by hydra from substrates produced by their endosymbiotic algae. *Nature* **221**: 381–382.

Sargent, M., and T. S. Austin. 1949. Organic productivity of an atoll. *Trans., American Geophysical Union* **30**: 245–249.

──── 1954. Biologic economy of coral reefs. U.S. Geological Survey, professional paper 260-E, pp. 299–300.

Slobodkin, L.B. 1962. Energy in animal ecology. In *Advances in ecological research,* vol. 1, J.B. Cragg, ed., pp. 69–101. New York: Academic Press.

Smith, D. C., L. Muscatine, and D. Lewis. 1969. Carbohydrate movement from autotrophs to heterotrophs in parasitic and mutualistic symbiosis. *Biological Reviews, Cambridge Philosophical Society* **44**: 17-90.
Stephens, G. 1960. Uptake of glucose from solution by the solitary coral, *Fungia*. *Science* **131**: 1532.
Stephenson, T. A. 1910. Coelenterata. I. *Actiniaria*. *British Antarctic Expedition* **5**: 1-68.
Stiven, A. E. 1965. The relationship between size, budding rate, and growth efficiency in three species of hydra. *Researches on Population Ecology* **7**: 1-15.
Taylor, D. L. 1968. *In situ* studies on the cytochemistry and ultrastructure of a symbiotic marine dinoflagellate. *J. Marine Biological Association, United Kingdom* **48**: 349-366.
———. 1969. The nutritional relationship of *Anemonia sulcata* (Pennant) and its dinoflagellate symbiont. *J. Cell Science* **4**: 751-762.
Trench, R. K. 1968. Liberation of soluble photosynthate by symbiotic zooxanthellae. *American Zoologist* **8**: 771.
———. 1969. The physiology and biochemistry of zooxanthellae symbiotic with marine coelenterates. Ph.D. dissertation, University of California at Los Angeles.
Wainwright, S. A. 1967. Diurnal activity of hermatypic gorgonians. *Nature* **216**: 1041.
Whitney, D. D. 1907. Artificial removal of the green bodies of *Hydra viridis*. *Biological Bulletin, Woods Hole* **13**: 291-299.
Yonge, C.M. 1957. Symbiosis. *Memoirs. Geological Society of America* **1 (67)**: 429-442.
———. 1963. The biology of coral reefs. In *Advances in marine biology*, vol. 1, F. S. Russell, ed., pp. 209-260. New York: Academic Press.
Yonge, C. M., and A. G. Nicholls. 1931. Studies on the physiology of corals. V. The effect of starvation in light and in darkness on the relationship between corals and zooxanthellae. *Scientific Reports, Great Barrier Reef Expedition* **1**: 177-211.
Zahl, P., and J. J. A. McLaughlin. 1959. Studies in marine biology. IV. On the role of algal cells in the tissues of marine invertebrates. *J. Protozoology* **6**: 344-352.

ALINA M. SZMANT CHAPTER 18
Scripps Institution of Oceanography, La Jolla, California

Patterns of $^{14}CO_2$ Uptake by *Chlorohydra viridissima*

The symbiotic relationship between algae and their coelenterate hosts has evoked much interest and speculation. Recent investigations have shown that the host tissues from some coelenterates receive either maltose (Muscatine 1965) or glycerol (Muscatine 1967) from their endosymbiotic algae. Yet little is known of how algae benefit from the relationship. It has been suggested that the host shelters the algae from excessive light, and that the algae obtain certain inorganic nutrients (such as phosphates and nitrates) and CO_2 from the host tissues (Yonge 1958). There is, however, no direct experimental evidence showing the extent of the host's contribution of CO_2 to the algae.

Previously Muscatine and Hand (1958), Lenhoff and Zimmerman (1959), and Goreau and Goreau (1960) showed that endosymbiotic algae photosynthetically incorporate $^{14}CO_2$ which is present in the environment of their coelenterate host. Other experiments with *Hydra littoralis* showed that, following the ingestion of food, the animals increased their production of CO_2 (Lenhoff and Loomis 1957). One might expect, therefore, that when well-fed specimens of *Chlorohydra viridissima* are supplied with $^{14}CO_2$, their algae would incorporate less of the label than would the algae from the starved hydra, because the $^{14}CO_2$ is diluted by unlabeled CO_2 derived from the metabolism of the host. On the other hand, nutrients taken up by the algal symbionts from the food digested by the host tissues might stimulate photosynthesis; hence, the rate of photosynthesis may exceed the rate of production of CO_2 by host tissues, thereby causing a net increase in $^{14}CO_2$ taken up from the environment.

In this paper, I describe relatively rapid and simple methods for directly demonstrating the uptake of $^{14}CO_2$ from the environment by algae in specimens of *Chlorohydra viridissima*. The symbiotic algae in this animal exhibited complex patterns of CO_2 fixation which were greatly affected by the nutritional state of the host.

MATERIALS AND METHODS

The green hydra, *Chlorohydra viridissima* (Florida strain, 1961), was grown in mass cultures (Loomis and Lenhoff 1956) in synthetic "M" culture solution (Muscatine and Lenhoff 1965). The cultures were kept at room temperature and 6 inches from a fluorescent bulb. The stock animals were fed once daily with freshly hatched *Artemia* nauplii. Only hydra with one small bud were used in the experiments.

The $^{14}CO_2$ was presented to the hydra in one of two ways: (*a*) groups of 5–10 hydra were placed in 3 ml of "M" solution in a sealed 7-ml Warburg flask lacking a center well; gaseous $^{14}CO_2$ was evolved by placing a solution containing 5 µc of $Na_2^{14}CO_3$ in the sidearm and then adding a few drops of phosphoric acid to that solution (Lenhoff 1959); (*b*) ^{14}C-sodium carbonate was added directly to the solution bathing the hydra in sealed 10-ml Erlenmeyer flasks. Both methods gave similar results. Incubation periods varied from 1 to 3 hours, depending upon the experiment.

At the end of the experiment the animals were washed serially three times with "M" solution, plated on Millipore filters (0.45-micron pore size), dried with an infrared lamp (Lenhoff 1959), and counted for radioactivity using a Nuclear Chicago gas flow counter, model 470.

For each point, groups of five hydra were used. They were plated together and counted together, and the resulting counts were divided by five. The protein content (Lowry et al. 1951) of groups of five hydra had a standard deviation of ± 10 percent.

RESULTS

$^{14}CO_2$ Uptake by *Chlorohydra viridissima*

The fixation of $^{14}CO_2$ by green hydra in the light is due mostly to the photosynthetic activities of the algae (Table 1). Animals with algae in

TABLE 1 Photosynthetic fixation of $^{14}CO_2$ by algae in *Chlorohydra viridissima*

Experimental Conditions	cpm/hydra
With Algae, Green Hydra	
1 day without food, incubated in light	3,302
1 day without food, incubated in dark	132
Without Algae, Albino Hydra	
1 day without food, incubated in light	60

NOTE: Groups of five hydras each were incubated in Warburg flasks for 3 hours with 5µc of $^{14}CO_2$ generated from the sidearm.

the dark and the aposymbiotic albino animals in the light fixed little $^{14}CO_2$.

Under the experimental conditions described, the $^{14}CO_2$ was fixed at a near constant rate for about 2 hours, usually leveling off after 2–3 hours (Figure 1). Therefore in this study we used incubation periods of either 2 or 3 hours, obtaining similar results in both cases.

Variability in the Effect of Food on $^{14}CO_2$ Uptake

To determine how the ingested food would affect $^{14}CO_2$ uptake by green hydra, groups of animals were fasted for periods varying from 10 to 72 hours. Half of each group was then fed to repletion with *Artemia* nauplii, and all groups were incubated for 3 hours with $NaH^{14}CO_3$. The results (Figure 2) showed a range of inhibitory to stimulatory effects,

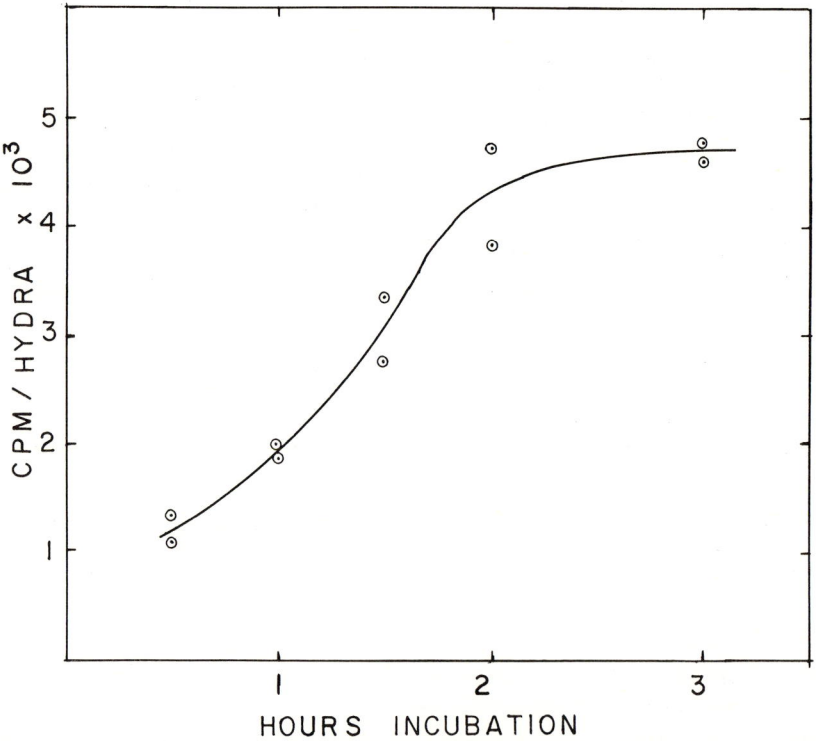

FIG. 1. Rate of $^{14}CO_2$ uptake by the algae in *Chlorohydra viridissima*. Groups of 10 hydra were incubated with 5 μc $Na_2{}^{14}CO_3$ for different periods of time. Points represent two 5-hydra samples, each of which was counted separately and the resulting cpm divided by 5.

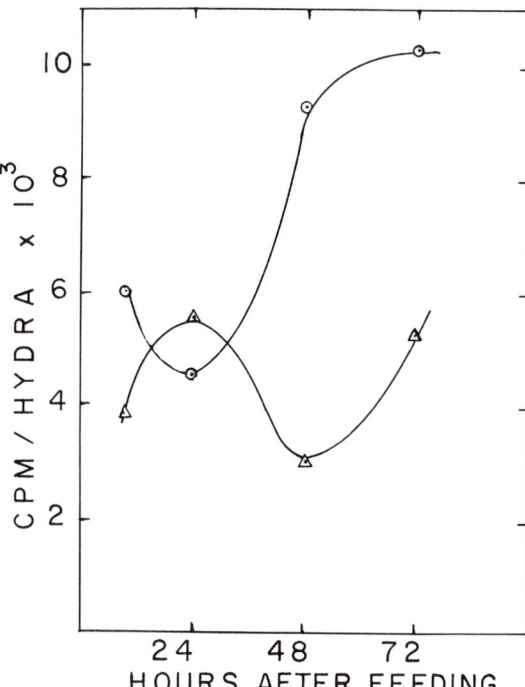

FIG. 2. Effect of food on the $^{14}CO_2$ uptake of fasted hydra. All animals were starved for the indicated time. Triangles represent experiments in which animals were fed 1 hour before incubation time; circles represent hydra incubated without prior feeding.

depending on the interval between the previous meal (zero time) and the time of the experiment.

To obtain a better picture of the degree of stimulation or inhibition of $^{14}CO_2$ fixation caused by the ingestion of food by hydra after they had been starved for various periods, we conducted the following experiments. Animals were fasted from 1 to 7 days. At daily intervals groups of 10 animals were divided into two groups, one of which was fed with *Artemia*. After 1 hour both groups were exposed to $^{14}CO_2$ for 3 hours. Both the fed and the starved animals followed similar patterns of fixation, the most activity occurring 2–3 days following the previous meal and declining steadily thereafter (Figure 3). Looking at the results from the fasted hydra, the lowered fixation no doubt reflects the progressively smaller sizes of the hydra as they were fasted for longer periods of time, and thus, possibly, contained fewer algae. On the other hand, the increase in $^{14}CO_2$ fixation between 1 and 2 days probably reflects the increase in hydra tissue and algae that occurs in the 2 days following a meal. Harder to evaluate are the effects of food on these fasted animals. It appears

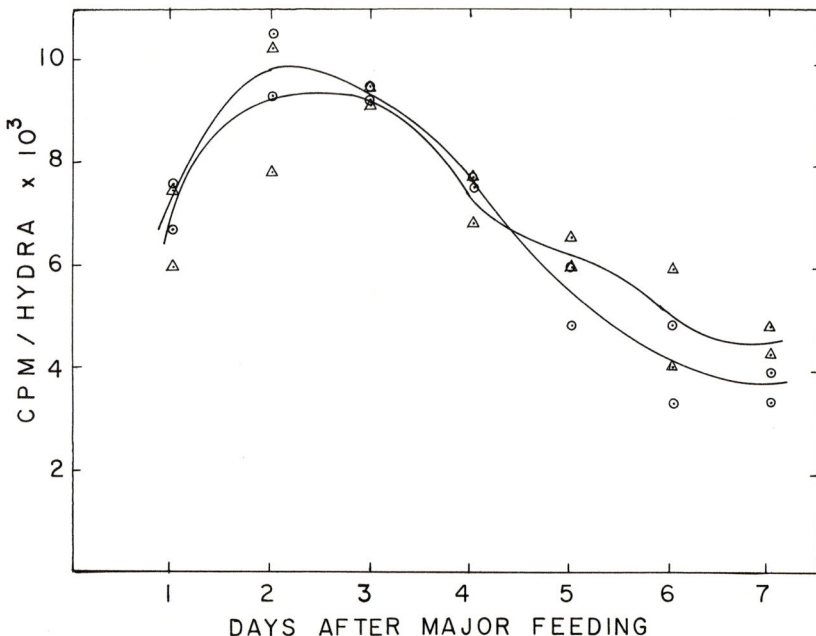

FIG. 3. Effect of food on the $^{14}CO_2$ uptake of fasted hydra. All animals were fed at time zero. Triangles represent experiments in which the animals were given a second feeding 1 hour before incubation time; circles represent hydra that were incubated without a second feeding. Prepared in collaboration with H. M. Lenhoff.

somewhat that the ingestion of food by hydra fasted for 1–2 days slightly decreases their $^{14}CO_2$ uptake; whereas the ingestion of food by hydra fasted 5–7 days slightly stimulates their $^{14}CO_2$ uptake.

Variation in $^{14}CO_2$ Fixation in Fasted Hydra

Restricting myself to the fasting period when the hydra did not show reduction in size, I carried out a series of 3-hour incubations. The animals used for the data in Figure 4A were fed once daily before experiment, while those in Figure 4B were fed on a 12-hour regime for 2 days before the experiment. For experiment 4A, the animals were fed as a single group and were removed for incubation at the desired intervals. For experiment 4B, separate groups of animals were fed over a staggered period so that all were ready for incubation at about the same time. The data show that a significant burst of $^{14}CO_2$ fixing activity occurred at about 16–20 hours. Minor peaks of activity also appeared to occur before and after the major one.

To get a better view of the patterns of $^{14}CO_2$ fixation in animals

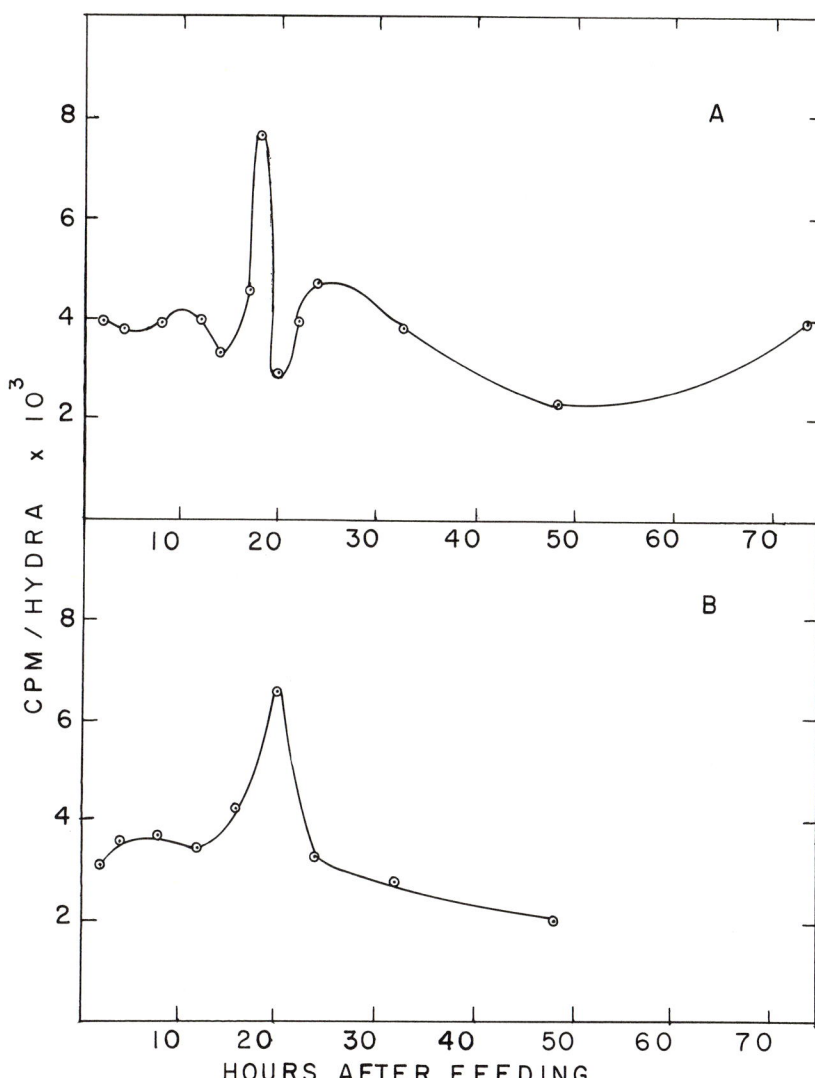

FIG. 4. Uptake of $^{14}CO_2$ at various times after feeding. Groups of 10 hydra were incubated with 5 μc $Na_2^{14}CO_3$ for 3 hours at various intervals after they were fed. Graph A represents experiments in which hydra were fed on a 24-hour schedule prior to the experiment. Graph B represents experiments in which hydra were fed on a 12-hour schedule for 2 days prior to the experiment.

starved up to 2½ days, we sampled the animals after random times of fasting and then incubated them with $^{14}CO_2$ for 1, 2, or 3 hours. The results (Figure 5) showed that regardless of the time of incubation all groups followed the same general pattern; that is, there were two large peaks of activity at about 17 and 40 hours, with the second one being slightly higher than the first. There was also a smaller peak of activity at 6 hours, but it was not always possible to detect this one in other experiments.

Although the method used in these experiments was adequate in surveying for patterns of $^{14}CO_2$ fixation by intact *Chlorohydra viridissima* and their endosymbiotic algae, it was not exact enough to permit fine analysis of these patterns. Any variation in size of animals selected, their algae, or their previous history could cause variations in the amount of $^{14}CO_2$ fixed. Further, variation might be caused by overlapping of the incubation periods with periods in which the CO_2 fixation patterns were changing. Despite these drawbacks similar patterns were found when experiments such as those described in Figure 5 were repeated, with the experimental differences being that the incubations were for 2 hours at 2-hour intervals, and the animals were controlled for previous exposure to light. In these experiments there usually appeared: (*a*) a decrease in activity during the first 3–6 hours after feeding; (*b*) a minor peak of activity at about 6–10 hours; (*c*) two major peaks of activity, one at 14–22 hours and the other at 36–48 hours; and (*d*) a decrease in activity following the second major peak.

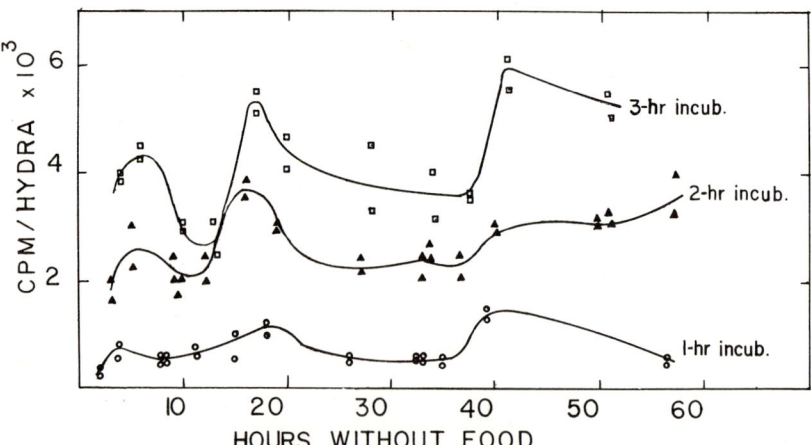

FIG. 5. Fixation of $^{14}CO_2$ at various times after feeding. The animals were presented with gaseous $^{14}CO_2$. Squares represent 3-hour incubation periods; triangles represent 2-hour incubation periods; and circles represent 1-hour incubation periods. Prepared in collaboration with H. M. Lenhoff.

DISCUSSION

This study indicates that the interactions between *Chlorohydra viridissima* and their associated algae are more complex than hitherto believed. The large variations in the amount of $^{14}CO_2$ taken up from the aqueous environment by the endosymbiotic algae *in vivo* suggest that the algae are heavily influenced by the nutritional state of the host. It seems reasonable that some of the released metabolites from the food stimulate algal photosynthesis, either by directly affecting the photosynthetic machinery of the algae or by providing substrates that funnel off the photosynthetic products as rapidly as they are formed.

A second way in which food could affect the amount of $^{14}CO_2$ fixed is by increasing the number of algae cells. Campbell (1967) demonstrated that the gastrodermal cells in *Hydra littoralis* go through stages of increased division at about 10 hours after the animals are fed. If the same holds true for *Chlorohydra viridissima*, we would expect that algal cell division would follow closely. Such a possibility might be checked by algal cell counts or by chlorophyll analysis.

Although the above may explain the observed increases in $^{14}CO_2$ fixation by the algae, it does not account for the decrease in uptake that usually immediately follows any increase. Perhaps the hydra cells inhibit the algal photosynthesis during those periods. Such control would prevent the alga-hydra relationship from becoming a parasitic one in which the algae kill their host. Hence, the interweaving of various processes that stimulates and inhibit algal photosynthesis might be expressed in some sort of biological rhythm (hinted at in Figure 5) in which the hydra, algae, or both are involved.

The variations in the uptake of environmental $^{14}CO_2$ by the algae *in vivo* (especially the inhibitions) might also be explained by fluctuations in the production of CO_2 evolved by the hydra's metabolism. Such an explanation would assume that the algal photosynthesis is more or less constant while the $^{14}CO_2$ taken into cells is diluted by $^{12}CO_2$ evolved from the host's metabolism. Food, naturally, would increase the production of $^{12}CO_2$ by the hydra (Lenhoff and Loomis 1957) and thus decrease the amount of $^{14}CO_2$ fixation. To determine if this view is correct, however, would require a careful kinetic analysis of the CO_2 production that precedes and follows the ingestion of food by aposymbiotic hydra.

Actually, the experiments show that the $^{12}CO_2$ produced by the hydra after they were fed did not cause a major decrease in the amount of $^{14}CO_2$ fixed. Perhaps the amount of such $^{12}CO_2$ may be small in comparison with the $^{14}CO_2$ pool supplied from the environment.

Other experimental approaches for determining the extent of CO_2 contributed by the hydra's tissues to their algae are possible. A promising method is evident from the work of Pearse (this volume). She studied calcification in corals by feeding them ^{14}C-labeled food and then measuring for radioactivity in the $CaCO_3$ skeleton. Similarly, it would seem profitable to feed green hydra on ^{14}C-labeled food to determine the amount of metabolic $^{14}CO_2$ produced by the hydra that is later taken up by the algae. In this sort of experiment, however, it would be necessary to distinguish between the fixation of $^{14}CO_2$ produced by hydra metabolism, and the ^{14}C-labeled metabolites taken up by the algae from their host's cytoplasm.

In conclusion, the experiments described in this paper show that the nutritional state of the hydra greatly affects photosynthesis in their algal endosymbionts. It is important to determine the nature of these effects if we are to fully understand alga-animal symbiosis, especially, in the case of corals, how the endosymbiotic algae affect calcification.

SUMMARY

1. The extent to which the algae of *Chlorohydra viridissima* utilize the metabolic CO_2 of their host was examined by comparing the amount of environmental $^{14}CO_2$ fixed by fed and starved hydra.

2. How the ingestion of food affected $^{14}CO_2$ uptake appeared to depend upon the interval between the previous meal and the time of the experiment.

3. A major increase in the amount of $^{14}CO_2$ fixed usually occurred at about 15-20 hours and around 40 hours after feeding.

4. Interpretations of these results and suggested experiments for investigating these phenomena further are presented.

LITERATURE CITED

Campbell, R. D. 1967. Tissue dynamics of steady state growth in *Hydra littoralis*. I. Patterns of cell division. *Developmental Biology* 15: 487-502.

Goreau, T. F., and N. I. Goreau. 1960. Distribution of labeled carbon in reef building corals with and without zooxanthellae. *Science* 131: 668-669.

Lenhoff, H. M. 1959. Migration of ^{14}C-labeled cnidoblasts. *Experimental Cell Research* 17: 570-573.

Lenhoff, H.M., and W.F. Loomis. 1957. Environmental factors controlling respiration in hydra. *J. Experimental Zoology* 134: 171-182.

Lenhoff, H. M., and K. F. Zimmerman. 1959. Biochemical studies of symbiosis in *Chlorohydra viridissima*. *Anatomical Record* 134: 599.

Loomis, W. F., and H. M. Lenhoff. 1956. Growth and sexual differentiation of hydra in mass culture. *J. Experimental Zoology* 132: 555-574.

Lowry, O. H., N. J. Rosebrough, A. L. Farr, and R. J. Randall. 1951. Protein measurement with the Folin phenol reagent. *J. Biological Chemistry* 193: 265-275.

Muscatine, L. 1965. Symbiosis of hydra and algae. III. Extracellular products of the algae. *Comparative Biochemistry and Physiology* **16**: 77–92.

———. 1967. Glycerol excretion by symbiotic algae from corals and *Tridacna* and its control by the host. *Science* **156**: 516–519.

Muscatine, L., and C. Hand. 1958. Direct evidence for the transfer of materials from symbiotic algae to the tissues of a coelenterate. *Proceedings, National Academy of Sciences* **44**: 1259–1263.

Muscatine, L., and H. M. Lenhoff. 1965. Symbiosis of hydra and algae. I. Effects of some environmental cations on growth of symbiotic and aposymbiotic hydra. *Biological Bulletin, Woods Hole* **128**: 415–424.

Yonge, C.M. 1958. Ecology and physiology of reef building corals. In *Perspectives in marine biology*, A. Buzzati-Traverso, ed., pp. 117–136. Berkeley and Los Angeles: University of California Press.

ERIC EISENSTADT
Washington University, St. Louis, Missouri

Transfer of Photosynthetic Products from Symbiotic Algae to Animal Tissue in *Chlorohydra viridissima*

The transfer of ^{14}C-labeled compounds from symbiotic algae to the tissue of their animal host has been demonstrated in some marine and freshwater coelenterates (Muscatine and Hand 1958; Goreau and Goreau 1960; Lenhoff and Zimmerman 1959; Muscatine and Lenhoff 1963). In corals the major carbon substance transferred from the zooxanthellae to host tissue is glycerol (Muscatine 1967), and in *Chlorohydra viridissima* the primary substance transferred is maltose (Muscatine 1965).

Using *Chlorohydra viridissima* Muscatine and Lenhoff (1963) showed that 12 percent of the radioactivity fixed by algal photosynthesis was transferred to the algae-free ectodermal tissue. They did not determine, however, the extent of transfer of label to the gastrodermal cells which house the algae.

The primary purpose of my study was to devise a method for complete separation of the symbiotic algae from the animal tissue of *Chlorohydra viridissima* to determine more precisely the extent of transfer of photosynthetically fixed ^{14}C-label from the algae to the host. Preliminary results on the effect of fasting on this transfer are also presented.

MATERIALS AND METHODS

All experiments were carried out with asexually reproducing *Chlorohydra viridissima* (Florida strain). The animals were maintained by mass culture methods (Loomis and Lenhoff 1956) in artifical "M" culture solution (Muscatine and Lenhoff 1965). Animals used for experiments were at a similar developmental stage, i.e., each animal had one bud.

Incubation Procedures

The animals were labeled by placing them in culture solution containing Na$_2$^{14}CO$_3$ in a 10-ml Ehrlenmeyer flask. For each experiment described in Table 1, 25 hydra were placed in 1 ml of "M" solution containing 10 μc Na$_2$^{14}CO$_3$. The flasks were incubated in a 30° C water bath illuminated by a 40-watt Sylvania Lifeline fluorescent tube placed 6 inches from the bath. At the end of the incubation period, the hydra were removed from the labeled solution and washed four times in clean culture solution.

Fractionation Procedures

The hydra were fractionated into the animal and algal components by the following method: Twenty-five hydra were gently homogenized in 0.5 ml "M" solution with a glass tissue grinder. The homogenate was centrifuged in a 3-ml conical centrifuge tube for 20 seconds in a clinical centrifuge set at ¾ maximum speed (Aloe Scientific Company, catalog number VE-428). The turbid white supernatant was removed from the dark green pellet, and the pellet was resuspended in 0.2 ml of culture solution and centrifuged again. The second supernatant was combined with the first and the volume was brought up to 1 ml. The combined supernatants are referred to as the *animal fraction*, and the green pellet as the *algal fraction*.

Radioactivity in the animal fraction was assayed by drying a 100-μl portion on an aluminum planchet, acidifying with 0.1 ml of 0.1 N HCl to evolve unincorporated $^{14}CO_2$, and by counting the planchets on a Nuclear-Chicago gas flow counter. Replicate counts were corrected for background radioactivity, and averaged. The algal pellet was prepared by resuspending and homogenizing with warm (80° C) 80 percent ethanol for 3 minutes. The volume was adjusted to 1 ml, of which a 100-μl portion was counted. In the rate experiment (Table 1), the pellet was resuspended in culture solution and then trapped on a Millipore filter and counted. Protein was determined by the method of Lowry et al. (1951).

Correction for Contamination of Animal Fraction with Radioactive Algal Pellet

To determine the extent to which algal material may have contaminated the animal fraction, I performed the following experiments: Algae, isolated from 25 hydra, were incubated in 10μc of $Na_2^{14}CO_3$ for 1 hour. The algae were then centrifuged, washed free of $Na_2^{14}CO_3$, and were gently homogenized with 25 aposymbiotic *Chlorohydra viridissima*. The homogenate was then separated into animal and algal fractions as described above, and samples of each fraction were counted. I found that the animal fraction always contained 10 percent of the ^{14}C originally incorporated by the algae. Therefore, in all the experiments described in the "Results" section, I have subtracted 10 percent of the radioactivity from the animal fraction and have increased the algal fraction accordingly. I did not run any controls to correct for animal nonphotosynthetic incorporation of $^{14}CO_2$, because Muscatine and Lenhoff (1963) showed this amount to be insignificant compared to the relative amount of

TABLE 1 Rates of incorporation of $^{14}CO_2$ by the algae of *Chlorohydra viridissima* and translocation of $^{14}CO_2$ to the animal tissues.

Incubation time	Radioactivity					
	Whole Hydra	Algae		Animal Tissue		
(hrs)	(cpm)	(cpm)	(cpm/µg protein N)	(cpm)	(cpm/µg protein N)	
1	32,500	19,000	2,400	13,500	1,780	
3	118,000	69,000	6,200	49,000	6,050	
5	152,000	77,000	9,800	75,000	8,150	

radioactivity fixed photosynthetically and transferred to the algae-free ectoderm during these short incubation periods.

RESULTS

Rate of Incorporation of $^{14}CO_2$ by Algae and Transfer of Radioactivity to the Host Animal Tissue

These rates were determined by incubating 200 hydra (fed 12 hours previously) in 5 ml of culture solution containing 25 μc of $Na_2{}^{14}CO_3$. After 1, 3, and 5 hours, 25 hydra were removed and separated into animal and algal fractions. The proteins of the animal and algal fractions were measured, and samples of the two fractions were counted for radioactivity.

Table 1 shows that the rates of incorporation of ^{14}C into the animal and algal fractions were very similar for the duration of the experiment. The animal fraction has 40 percent of the total label after 1 hour and 3 hours of incubation and 50 percent after 5 hours of incubation.

The results of the same experiment show that the specific activity (cpm/μg protein N), of the animal fraction ranged from 74 percent to 98 percent of the values in the algal fraction.

Effect of Fasting on the Relative Specific Activity of Label in the Animal Fraction

Table 2 gives data on the amount of label and specific activity of that label found in the animal fraction of hydra fasted 11, 34, 36, or 60 hours before being exposed to $Na_2{}^{14}CO_3$. In each experiment, 25 hydra were incubated in 10 μc $Na_2{}^{14}CO_3$ in 1-ml culture solutions for 2 hours. The results show that both the amount of radioactivity and the specific activity of the animal tissue from hydra fasted 11 hours were about 3 to 4 times higher than the similar values for hydra fasted 34 to 60 hours before being exposed to the $Na_2{}^{14}CO_3$.

DISCUSSION

These results show that the endosymbiotic algae of *Chlorohydra viridissima* contribute to the animal tissues 30–40 percent of the label which they have incorporated from exogenous $^{14}CO_2$. This process requires about 2 hours. Extrapolating from the data of Muscatine and Lenhoff (1963), who showed that the algae-free ectoderm received about 12 percent of the label in 2 hours, I calculated the level of radioactivity at 2–3 hours to be distributed with 65 percent of the label in the algae, 25 percent in the endoderm (without algae) and 10 percent in the ectoderm.

Hence, my rate experiments (Tables 1 and 2) showed that the endosymbiotic algae transfer a significant amount of their photosynthetic

TABLE 2 Effects of fasting on amount of $^{14}CO_2$ fixed, the percent of the label transferred, and the specific activity of label in the animal tissue.

Time Since Hydras' Last Meal (hrs)	Total ^{14}C Incorporated (cpm)	Activity in Animal Fraction (cpm)	(%)	Protein Nitrogen (µg)	Specific Activity Animal Fraction (cpm/µg)
11	159,154	50,460	31.7	3.4	14,841
	184,831	43,990	23.8	2.0	21,995
34	67,564	30,274	44.8	6.4	4,730
	38,755	16,500	42.6	5.0	3,300
36	74,389	11,532	15.5	2.8	4,118
	53,093	14,590	27.5	1.6	9,119
60	52,601	14,170	26.9	3.4	4,168
	40,596	13,137	32.4	2.2	5,971

NOTE: 25 hydra were incubated in 10µc $Na_2^{14}CO_3$ in 1-ml culture solution for 2 hours.

products to the animal tissue, and that they do this rapidly at a near constant rate. The data show that the specific activity of the label in the animal fraction (cpm/µg tissue N) is approximately that of the algal fraction. To determine the metabolic fate of the labeled material in the plant and animal tissues, it would be possible to place such labeled hydra into a medium with unlabeled Na_2CO_3 for varying periods of time, and then determine the subsequent distribution of label.

The work of Szmant (this volume) shows that the amount of $^{14}CO_2$ fixed by the algae of *Chlorohydra viridissima* can vary significantly depending upon the interval between the hydra's last meal and the incubation experiment. Thus, we might expect the percentage of the label transferred to the animal tissues to be also dependent upon the time of the hydra's last feeding. The data in Table 2 show that the greatest amount of $^{14}CO_2$ was incorporated if the incubation was done 11 hours after the hydra was fed, and hence, agree with the results of Szmant's experiments. The data are not sufficiently conclusive, however, to show that prolonged fasting significantly affects the percentage of label transferred to the animal tissues. The specific activity of radioactivity in the animal fraction, like the amount of $^{14}CO_2$ fixed, shows a similar quantitative relationship to the time at which the hydra received its last meal.

SUMMARY

1. By allowing *Chlorohydra viridissima* to fix $^{14}CO_2$, and then separating endosymbiotic algae from its tissues, we were able to show that algae contribute to the animal tissues 30–40 percent of the ^{14}C-label incorporated by algal photosynthesis.

2. The specific activity of the label transferred to the animal tissues was approximately that of the algae.

3. The amount of label transferred was not significantly affected by the degree to which the hydra were fasted.

LITERATURE CITED

Goreau, T. F., and N. I. Goreau. 1960. Distribution of labeled carbon in reef-building corals with and without zooxanthellae. *Science* **131**: 668–669.

Lenhoff, H. M., and K. F. Zimmerman. 1959. Biochemical studies of symbiosis in *Chlorohydra viridissima. Anatomical Record* **134**: 599.

Loomis, W. F., and H. M. Lenhoff. 1956. Growth and sexual differentiation of hydra in mass culture. *J. Experimental Zoology* **132**: 555–574.

Lowry, O. H., N. J. Rosebrough, A. L. Farr, and R. J. Randall. 1951. Protein measurement with the Folin phenol reagent. *J. Biological Chemistry* **193**: 265–275.

Muscatine, L. 1965. Symbiosis of hydra and algae. III. Extracellular products of the algae. *Comparative Biochemistry and Physiology* **16**: 77–92.

———. 1967. Glycerol excretion by symbiotic algae from corals and *Tridacna* and its control by the host. *Science* **156**: 516–519.

Muscatine, L., and C. Hand. 1958. Direct evidence for the transfer of materials from symbiotic algae to the tissues of a coelenterate. *Proceedings, National Academy of Sciences* **44**: 1259–1263.

Muscatine, L., and H. M. Lenhoff. 1963. Symbiosis: on the role of algae symbiotic with hydra. *Science* **142**: 956-958.

———. 1965. Symbiosis of hydra and algae. I. Effects of some environmental cations on growth of symbiotic and aposymbiotic hydra. *Biological Bulletin, Woods Hole* **128**: 415–424.

AMADA REIMER
University of Southern California, Los Angeles

CHAPTER 20

Uptake and Utilization of ^{14}C-Glycine by *Zoanthus* and Its Coelenteric Bacteria

The ability to remove amino acids and other small organic compounds from dilute solutions is widespread among marine invertebrates. Stephens and Schinske (1961) tested specimens from 10 phyla of marine invertebrates and concluded that some soft-bodied marine invertebrates, when exposed to an amino acid such as glycine or phenylalanine at concentrations between 10^{-5} and 10^{-6} moles per liter, have the capacity to remove it quite rapidly from solution.

The question arose as to whether the removal of amino acids from solution might satisfy part of the nutritional requirements of the zoanthid *Zoanthus sandwichensis*. Unlike the closely related *Palythoa* sp., *Zoanthus* can not be made to feed on brine shrimp (*Artemia salina*) nauplii, nor will *Zoanthus* respond to either shrimp or oyster extracts. In fact, nothing is known about the feeding habits of this organism. This paper describes the uptake of ^{14}C-glycine solution by *Zoanthus sandwichensis* and by bacteria living in the animal's coelenteron, and the metabolic fate of glycine in these organisms.

MATERIALS AND METHODS

Collection and Maintenance of Organisms

Zoanthus sandwichensis (Walsh and Bowers) was collected in Kaneohe Bay, Oahu, Hawaii, and maintained in tables of running seawater. For each experiment, animals were chosen from the same colony, were detached from the substrate, cleaned of calcareous and epiphytic materials, and kept in finger bowls containing artificial seawater ("Instant Ocean," Aquarium Systems, Inc.) for 1 day. The water was replaced with clean seawater 1 hour prior to an experiment.

Incubation with ^{14}C-glycine

^{14}C-glycine was added to a beaker of 100 ml artificial seawater to give a final concentration of 2.16 mg/ml and 1 µc/ml. A cluster of five zoanthid polyps was added to the solution, and was then incubated at 28° ± 0.5° C in a plexiglass water bath above four 40-watt Sylvania fluorescent lights. Incident illumination was approximately 560 footcandles. After being incubated for 2–13 hours, the animals were removed, rinsed in 14–17 brief serial changes of artifical seawater, and analyzed as described below.

Fractionation of Labeled Zoanthids

Labeled zoanthids were homogenized with a mortar and pestle to a fine suspension in 95 percent ethanol. The suspension was washed into a test tube and the volume adjusted to 10 ml. A 0.1-ml portion was removed for protein determination (Lowry, et al. 1951). The suspension was centrifuged for 2 minutes at 1,500 rpm (International Clinical Centrifuge), and the alcoholic supernatant was withdrawn. The insoluble residue was further extracted with 0.5-ml portions of 80 percent ethanol, 50 percent ethanol, and finally with hot distilled water, until a gray-white residue was obtained. Supernatant and extracts were combined and stored at -15° C. Insoluble residue was hydrolyzed in a sealed tube with 6 N HCl at 108° C for 24 hours.

Chromatography

Ethanolic extracts and neutralized hydrolyzates were analyzed by two-dimensional chromatography on Whatman no. 4 paper (18½" × 22½") using phenol:water (72:28; w/w) in the first dimension and butanol:propionic acid:water (1246:620:884; v/v/v) in the second dimension. Papers were exposed to Kodak blue-sensitive "No-screen" medical X-ray film for 2–14 days (Benson et al. 1950).

Assay of Radioactivity

Fluid samples of known volume were dried on planchets and assayed with a thin end window Geiger tube or Nuclear-Chicago gas flow detector, model 470, with correction for background. Radioactive spots on paper chromatograms were assayed with an end window probe. Radioactivity of each spot was expressed as a percentage of the total activity detected on the paper.

Identification of Radioactive Unknowns

Tentative identifications were made by comparing radioautographs of chromatographed unknowns with the chromatographic maps of Bassham and Calvin (1957) by elution and co-chromatography with authentic labeled and nonradioactive compounds and, in the case of amino acids, by comparing ninhydrin (1 percent in acetone) positive spots with those of known amino acids.

Bacterial Analysis

Prior to each experiment, a sample of coelenterate fluid was removed and the number of bacteria in the sample determined directly with the aid of a hemocytometer. A sample of coelenteric fluid was introduced to 100 ml of sterile artificial seawater enriched with 0.5 gram Bactopeptone and 0.1 gram dextrose, and ^{14}C-glycine (1 μc ml; approx. 0.19 mg/ml). Growth was monitored daily by cell counts of samples. At the onset of the stationary phase (16–20 hours), the suspension was centrifuged for 30 minutes at 12,000 rpm (Servall Refrigerated Centrifuge). The bacterial pellet obtained was washed in artificial seawater and was fractionated and analyzed as described for the zoanthids.

RESULTS

Uptake of ^{14}C-glycine from Solution

Table 1 shows the time course of removal of ^{14}C-glycine from solution and its accumulation by the zoanthids. Of the total ^{14}C available, the animals acquired 65 percent after 13 hours. The relative specific activity (cpm/μg protein nitrogen) of the animals also increased appreciably during this time. The seawater in which the animals were rinsed after the 2-hour and 6-hour incubation periods contained 29 percent and 25 percent, respectively, of the total glycine. After 10–13 hours, however, only 5 percent remained in the wash. Whether or not the radioactivity in the wash represented material from the surface of the animals or from the coelenteric fluid, or efflux of intracellular glycine cannot be determined from the data acquired. Further, it was noted in one control experiment, consisting of the incubation mixture without the animals, that 4–7 percent of the initial ^{14}C-glycine added could not be recovered after 10 hours. This suggests that a fraction of the glycine is removed from solution by some means other than uptake by the animals.

Analysis of ^{14}C in Alcohol-Soluble Fraction of Zoanthids

Regardless of the length of incubation, about 90 percent of the total ^{14}C taken up by the zoanthids was always in the alcohol-soluble fraction. Radiochromatographic analysis of this fraction after 2, 6, and 10 hours revealed up to eight different labeled products. A representative radiochromatogram of the compounds which were detected is shown in Figure 1. The major radioactive compounds in this fraction were tentatively identified as ^{14}C-glycine and lipid. At least five other labeled compounds were detected, only one of which (number 3) exceeded 5 percent of the total radioactivity in the fraction. The others, which were present in small amounts, are grouped together in the last column of Table 2 for

TABLE 1 Uptake of ^{14}C-glycine from solution by Zoanthus

Incubation Time (hrs)	In Medium Initial (cpm)	Final (cpm)	In Animals (cpm)	%	In Wash (cpm)	%	Recovery %	Specific Activity (cpm/µg protein N)
2	1,365,841	803,244	210,640	15.4	399,560	29.2	103.1	432
6	1,472,040	249,424	614,080	41.5	368,090	25.0	83.5	787
10	1,148,690	450,520	544,880	47.5	49,710	4.3	91.0	865
13	1,860,960	182,880	1,214,656	65.0	86,907	4.6	79.4	1,084

TABLE 2 Distribution of ^{14}C in alcohol-soluble fraction of Zoanthus as determined by radiochromatography

Incubation Time (hrs)	As Glycine	Radioactivity (%) As Lipid	As Unknown no. 3	As unknown nos. 1, 2, 4, 5, 6
2	81.0	1.6	1.5	15.9 (no. 6 absent)
6	66.0	5.2	21.7	7.1
10	48.1	42.0	6.5	3.4 (nos. 1, 2, 4 absent)

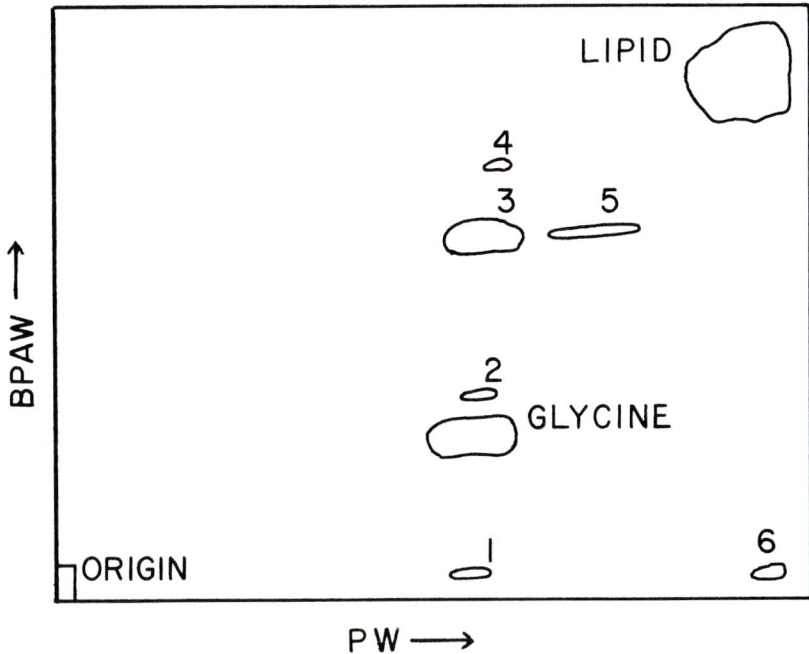

FIG. 1. Radiochromatogram of the alcohol-soluble fraction of *Zoanthus* after 6 hours' incubation with ^{14}C-glycine.

convenience. Table 2 shows that the relative amounts of these substances varied with incubation time. The amount of ^{14}C-glycine in the alcohol-soluble fraction decreased from 81 to 48 percent between the 2- and 10-hour incubations. There was a concomitant increase in the lipid fraction from 1.6 to 42 percent during this time, while compound number 3 increased to 21 percent but then decreased to 6.5 percent. It appears that under these conditions ^{14}C-glycine is utilized by zoanthids mainly in the synthesis of lipid. Compound number 3 may represent an intermediate in the pathway.

Uptake of ^{14}C-glycine by Coelentric Bacteria

The coelenteric fluid of *Zoanthus sandwichensis* contained diverse unicellular organisms. The most numerous of these was an unidentified species of bacterium, usually occurring as 1–2 micron motile rods that occasionally formed chains. The concentrations of these bacteria were two or three orders of magnitude higher than the concentrations usually found in marine waters (Zobell and Feltham 1938). To determine if this bacterial flora had been accumulating glycine from the solution to a significant extent, an inoculum of coelentric fluid was incubated with

^{14}C-glycine for 16–20 hours. Results are shown in Table 3. In two experiments, the cultured bacteria grew rapidly and removed 5.0 percent or less of the total labeled glycine from the medium. Table 4 shows distribution of radioactivity in the alcohol-soluble and alcohol-insoluble fractions of the bacteria as determined by chromatographic analysis. Almost 90 percent of the total ^{14}C in the alcohol-soluble fraction was associated with the lipid fraction. The remainder was nearly all unincorporated ^{14}C-glycine. The converse was observed in the hydrolyzed alcohol-insoluble fraction, probably protein; the major labeled constituent was ^{14}C-glycine while the labeled lipid accounted for only 15 percent of the total insoluble activity. Since the concentration of bacteria was higher in the cultures than in the coelenteric fluid, and the uptake of glycine only 5.0 percent or less, I concluded that activities of the bacteria *in situ* did not significantly affect the results of uptake of glycine by zoanthids. This conclusion is based on the assumption that coelenteric bacteria behave like those in culture.

DISCUSSION

The experiments reported here show that *Zoanthus sandwichensis* (Walsh and Bowers) accumulates ^{14}C-glycine from solution and incorporates a large proportion of the label into lipid. The zoanthid harbors a large bacterial flora which, in free culture, can also accumulate ^{14}C-glycine from solution but to a lesser extent than can the host tissues. The question remains, however, as to how these observations relate to the nutrition of *Zoanthus*.

Long and patient observations in the laboratory failed to give any clues as to how these zoanthids feed on particulate matter. If no conventional overt feeding behavior is displayed under natural conditions during the day or night, there are three possibilities which might explain how *Zoanthus* fulfills its nutritional needs: (1) That it can accumulate organic material from solution has been shown in this study. Before these data can be considered significant, it is important to know the concentration of glycine in the habitat of *Zoanthus* and the rate and extent of uptake of glycine at that concentration. If we average the values found by Belser (1959) for inshore temperate waters, and by Stephens (1963) for mud flat interstitial waters, we arrive at a figure of 0.750 milligrams glycine per milliliter. This is about one-third the concentration used in the present study. (2) The possibility that zoanthids can "farm" a large bacterial flora within the coelenteron and perhaps feed on them has not been tested, but such bacteria certainly represent a source of nutrients which would preclude overt feeding behavior. Some aspects of feeding on bacteria by corals are discussed by DiSalvo in Chapter 12. (3) A third possible source of nutrients for *Zoanthus* is their symbiotic zooxanthellae.

TABLE 3   Uptake of ^{14}C-glycine by bacteria after 16-hour incubation and growth

Initial Number of Bacteria ($\times 10^4$)	Final Number of Bacteria ($\times 10^4$)	Generation Time (hrs)	Initial ^{14}C in Medium (cpm)	^{14}C in Bacteria (cpm)	^{14}C incorporated (%)
375	34,250	4.5	170,830	6,894	3.5
130	2,950	4.5	37,930	2,268	5.0

TABLE 4 Distribution of ^{14}C in bacteria after 16-hour incubation and growth

Fraction	Radioactivity (%)		
	As Glycine	As Lipid	As Unknown
Alcohol-soluble	9.7	88.2	2.1
Alcohol-insoluble	79.6	14.9	5.1

It is now established that zooxanthellae, symbiotic with a wide range of marine organisms including corals and anemones, release organic material to their host tissues (Muscatine 1967). Evidence for assimilation of zooxanthellae photosynthate by some zoanthids is given by von Holt and von Holt (1968a, b).

These three factors together may account for the greater part of the organic nutrition of *Zoanthus*. For example, the uptake of small molecules from solution may provide the animal with nitrogen, as well as phosphorus, trace metals, and other inorganic nutrients. The zooxanthellae may provide reduced organic carbon. The bacteria, which feed on soluble substrates in their environment, may provide *Zoanthus* with complex organic materials such as vitamins. Thus, if *Zoanthus* does not feed on plankton in nature, its accumulation of amino acids from solution may be nutritionally significant.

SUMMARY

1. *Zoanthus sandwichensis* accumulated about 65 percent of the ^{14}C-glycine present in the seawater at a concentration of 2.16 mg/ml.

2. After 13 hours, a large proportion of the ^{14}C-glycine was metabolized, and the ^{14}C was utilized in synthesis of protein and lipid.

3. The animals harbored coelenteric bacterial floras which were cultured *in vitro*. The cultured bacteria accumulated some ^{14}C-glycine which they utilized in the synthesis of lipid and protein.

4. The possible importance of glycine uptake to nutrition of zoanthids is discussed.

LITERATURE CITED

Bassham, J. A., and M. Calvin. 1957. The path of carbon in photosynthesis. Englewood Cliffs, N.J.: Prentice-Hall.

Belser, W. 1959. Bioassay of organic micronutrients in the sea. *Proceedings, National Academy of Sciences* 45: 1533–1542.

Benson, A. A., J. A. Bassham, M. Calvin, T. C. Goodale, V. A. Haas, and W. Stepka. 1950. The path of carbon in photosynthesis. V. Paper chromatography and radioautography of the products. *J. American Chemical Society* 72: 1710–1718.

Holt, C. von, and M. von Holt. 1968a. Transfer of photosynthetic products from zooxanthellae to coelenterate hosts. *Comparative Biochemistry and Physiology* **24**: 73-81.

———. 1968b. The secretion of organic compounds by zooxanthellae isolated from various types of *Zoanthus. Comparative Biochemistry and Physiology* **24**: 83-92.

Lowry, O. H., N. J. Rosebrough, A. L. Fair, and R. J. Randall. 1951. Protein measurement with the Folin phenol reagent. *J. Biological Chemistry* **193**: 265-276.

Muscatine, L. 1967. Glycerol excretion by symbiotic algae from corals and *Tridacna* and its control by the host. *Science* **156**: 516-519.

Stephens, G. C. 1963. Uptake of organic material by aquatic invertebrates. II. Accumulation of amino acids by the bamboo worm, *Clymenella torquata. Comparative Biochemistry and Physiology* **10**: 191-202.

Stephens, G. C., and R. A. Shinske. 1961. Uptake of amino acids by marine invertebrates. *Limnology and Oceanography* **6**: 175-181.

Zobell, C. E., and C. B. Feltham. 1938. Bacteria as food for certain invertebrates. *J. Marine Research* **4**: 313-325.

CLAYTON B. COOK
Duke University, Durham, North Carolina

Transfer of ^{35}S-Labeled Material from Food Ingested by *Aiptasia* sp. to Its Endosymbiotic Zooxanthellae

Many of the biochemical studies on alga-invertebrate symbiosis have dealt with the contribution of the alga to the animal tissue. Little is known about the extent to which algae derive nutrients from the animal tissues in which they grow. We may assume that algae receive some CO_2 from animal metabolism, but they still need such substances as salts, metals, phosphorus, nitrogen, and sulfur. If it could be shown that an alga-animal association can survive without the animal taking in those nutrients with its food, then we might assume that the alga obtains the nutrients directly from the environment. Since there are no thoroughly investigated cases of a self-sufficient alga-animal relationship such as the one reported by Gohar (1940), it seems more likely that algae obtain their protein nitrogen and sulfur, for example, directly from the metabolic pools of the animals' tissues. To test this view, I fed mouse liver labeled with ^{35}S amino acids to symbiotized *Aiptasia* sp. Because *Aiptasia* can digest and assimilate mouse liver, any label found in its algal cells would necessarily have originated from the food ingested by the animal.

This report contains my preliminary results, a discussion of the pitfalls in this type of research, and suggestions regarding other approaches to problems of animal-to-alga transfer of materials.

MATERIALS AND METHODS

Specimens of *Aiptasia* sp. were collected from masses of the siphonaceous green alga *Dictyosphaeria intermedia* growing on reefs in Kaneohe Bay, Oahu, Hawaii. The animals were maintained without added food in running seawater for 2-5 days preceding their use.

Label was administered to the anemones by micropipetting into their coelenterons a suspension of ^{35}S-labeled mouse liver and unlabeled *Artemia salina* nauplii, both of which had been macerated in a glass tissue grinder. Immediately following this procedure, and once daily thereafter, the anemones were fed intact *Artemia* nauplii. The labeled animals in beakers containing 20-30 ml of seawater were kept in the laboratory under 24-hour fluorescent illumination. The medium was changed prior to each experiment and also at 24-hour intervals whenever a day or more elapsed between samplings.

The amount of activity in the zooxanthellae of an individual anemone was assayed at different times by the following technique: Several full-sized tentacles were removed from a labeled animal and drained on the side of the beaker. The tentacles, in 0.1 ml of Millipore-filtered seawater, were broken up by being drawn and expelled through a 1-ml syringe fitted with a 2-inch no. 23 needle until an even suspension was obtained. The resulting suspension was centrifuged at about 4,500 rpm for 30 seconds in an International Model CL Centrifuge. The supernatant was removed and the pellet washed three times with filtered seawater. The washes were combined with the original supernatant, and a sample of this fluid was placed on a planchet, dried, and counted. The pellet was resuspended in 0.2 ml of seawater, placed on a planchet, and counted. Microscopic examination of the pellet with a Leitz Labolux phase contrast microscope showed that it consisted largely of zooxanthellae and a small number of nematocysts.

For experiments determining uptake of label by the algae in the entire sea anemone (not just the algae of the tentacles), animals having expanded oral discs of 8–10 mm were used. Each animal was homogenized with 1.0 ml filtered seawater in a glass tissue grinder. In those few instances when the anemone produced excessive mucus, an additional 0.4 ml of seawater was added. Following homogenization, the volume was brought to 2.0 ml with seawater, and the suspension was centrifuged at 9,000 rpm for 30 seconds. In those tubes having excessive mucus, the centrifugation was repeated a second time. The supernatant, which was mostly animal tissue, was removed; its volume was adjusted to 2.0 ml, and 0.1 ml of this was placed on a planchet and counted. The pellet, which contained most of the algae, plus some mucus and bits of insoluble animal tissue, was washed three times with 1.0-ml portions of seawater. From the final suspension, 0.1 ml was placed on a planchet and counted.

The fate of alcohol-soluble ^{35}S material in animal and algal compounds was studied by the following method. Mouse liver was extracted first with 80 percent ethanol at 70° C, then with 100 percent ethanol at 70° C; this was followed by an extraction with 95 percent ethanol and 50 percent ethanol at room temperature. The combined alcohol extracts were evaporated to dryness and the residue of alcohol-soluble material was taken up with seawater. This material was micropipetted into the coelenterons of the anemones. The distribution of the label in the various chemical fractions of the animal tissue and algal cells was determined using a modified Schmidt-Thanhauser technique. To help break up the algae for the fractionation study, the algal pellet was washed and ground three times with carborundum particles; the carborundum was removed by centrifuging at 4,500 rpm for 10 seconds. It should be noted that the gut contents of the anemones were not flushed prior to the fractionation; this point has significant bearing on the results.

The radioactive mouse liver fed to the anemones was fractionated, and was found to contain 54.5 percent of its ^{35}S in the trichloroacetic-acid-soluble (TCA-sol) fraction, 18.1 percent in the TCA-soluble and alcohol-soluble (alc-sol) fraction, and the remaining 25.1 percent in the TCA-alcohol insoluble (alc-insol) fraction. Previous studies (Lenhoff 1961) showed the label in the TCA-sol fraction to be mostly in ^{35}S amino acids; in the alc-sol fraction, mostly in large peptides or small proteins; and in the alc-insol fraction, mostly in protein. The TCA-sol and alc-sol values are relatively higher than those previously reported for mouse liver labeled by this technique (Lenhoff 1961); however, see the data of Murdock in this volume.

RESULTS

Uptake of ^{35}S by Zooxanthellae Isolated from *Aiptasia* Tentacles

Figure 1 shows the results of experiments in which I measured the radioactivity found in the algae isolated from tentacles of *Aiptasia* at times varying from 1 to 96 hours after the labeled liver was ingested by the anemone. The data are expressed as a percentage of the total ^{35}S in both animal tissue and in the algae; each plot in Figure 1 represents the results obtained from one animal.

Despite the variability occurring among the individual experiments, some general conclusions can be made. (1) The maximum amount of label found in the algae during the assay period reached between 20–30 percent of the total radioactivity present in the tentacle. Assuming some contamination from labeled animal particles, 20 percent would be a safe estimate. (2) Labeling of the algae occurred within a few hours after the label was given to the anemone. Hence, future kinetic studies should be done at even earlier times. (3) There appear to have been periods after the anemones were fed at which the algae retained greater amounts of label, some animals (A and B, and possibly C and D) having two such peaks. (4) A steady state appears to have been reached (at least in A and B) in which 10–15 percent of the radioactivity was retained after 80 hours.

Uptake of ^{35}S by Zooxanthellae Isolated from Entire Animals

To determine whether the 10–15 percent of the label retained by the algae from tentacles after 80 hours (Figure 1, A and B) also applied to algae isolated from entire animals, four anemones were fed ^{35}S-labeled liver, and the animal and algal components were separated by maceration and centrifugation 5 days later. The percentages of the total activity

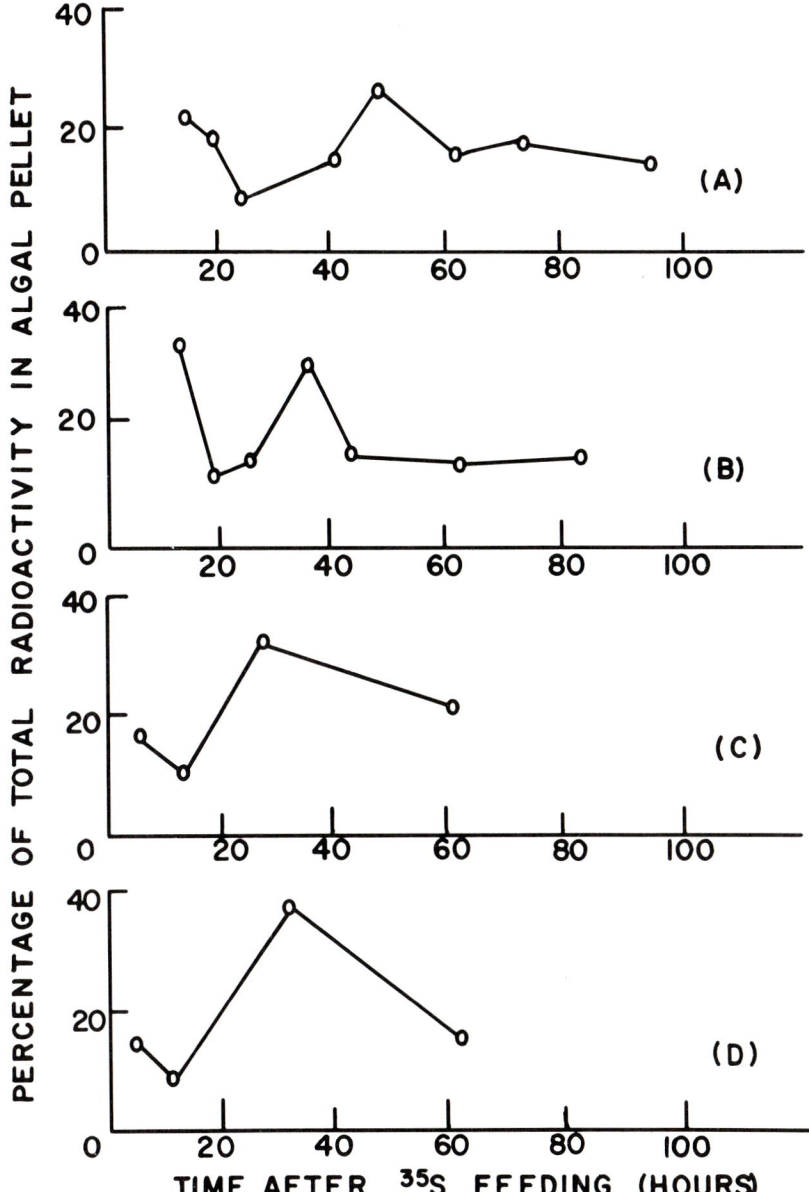

FIG. 1. The percentage of ^{35}S found in symbiotic zooxanthellae isolated from tentacles of *Aiptasia*. Each plot represents the results of an experiment in which a single animal was sampled at various times after it had ingested labeled food. Note that fluctuations occur in these values, and that a consistent value of 15–20 percent appears to have been reached 60 hours after the label was administered.

TABLE 1 Percentage of ^{35}S recovered from algal pellets of whole anemones 5 days after receiving label

Anemone No.	Activity in the Algal Pellet (%)
1	18.6
2	35.5
3	27.4
4	44.0

found in the algal pellets of these anemones are listed in Table 1. The mean of 30 percent retained was significantly greater than the 10–15 percent value found for algae isolated from tentacles. Clean algal separations were difficult to obtain by this method, due to the presence of insoluble animal tissue and mucus. (This tissue and mucus were not encountered in the tentacle-separation technique.) The higher percentages of activity in the whole animal pellets were probably the result of this contamination.

Distribution of Label into Algal and Animal Fractions

Examination of the data in Figure 1 suggests that the algae take up and exchange amino acids directly with the amino-acid pool of the animal tissue. Because (a) significant label was found in the algae soon after the anemones were given the labeled food, and (b) the percentage of label in the algae did not seem fixed, but rather varied with time, it was decided to test the ability of the algae to take up small molecules. To do this, some anemones were given an aqueous solution of the dried alcohol extract of the labeled mouse liver and fed *Artemia* nauplii immediately thereafter. After 18 hours the algal pellet was separated from the animal tissues. Both the algal pellet and the animal fraction were fractionated with TCA and ethanol. The results (Table 2) showed that both the algal and animal material had about the same distribution of radioactivity among the three fractions isolated. Most of the label was in the TCA-sol fraction while little ^{35}S was found in either the alc-sol or alc-insol fractions.

TABLE 2 Percentage of ^{35}S recovered from algal pellets and from tissues of anemones 18 hours after feeding

Fraction	Total Activity	
	Algal Pellet (%)	Animal Tissues (%)
TCA-soluble	90.6	92.2
Alcohol-soluble	4.3	4.7
TCA/Alcohol-insoluble	5.1	3.1

DISCUSSION

These preliminary results point out some of the difficulties in developing the experimental methodology for studying the contribution of the host animal tissue to the nutrition of the algal endosymbiont.

By feeding the anemone labeled food and subsequently analyzing the algae for radioactivity, I was able to show that the algae take up an appreciable amount of the label within a few hours (Figure 1), and retain about 15–30 percent of the label for at least 3–5 days (Figure 1; Table 1).

There remains the problem, however, of proving conclusively that all the radioactivity found in the so-called algal pellet is in the algae and not in bits of animal tissues that may have remained in the algal pellet after centrifugation. High resolution radioautography of a smear of the algal pellet should indicate the extent of contamination with labeled animal particles.

Perhaps this source of error could be bypassed by feeding the animal a label that could be incorporated into a specific plant material. For example, after giving the anemone ^{15}N-labeled food, the chlorophyll could be extracted, purified, and analyzed for ^{15}N. In such an experiment, any isotope found in the chlorophyll would be a true indication of a contribution to algal metabolism.

The experiments showing that the proportion of radioactivity in the algae varies with time (Figure 1) suggest that the ^{35}S-amino acids in the algal pellet are not fixed into protein, but rather are unbound and are free to interchange with the amino-acid pool of the animal tissue. Alternatively, the peaks in Figure 1 may represent periods of great algal activity. For example, dinoflagellates, the group with which zooxanthellae are classified (McLaughlin and Zahl 1959), exhibit persistent rhythms in photosynthetic capacity (Hastings, Astrachan, and Sweeney 1961). During periods of intensified algal activity, more ^{35}S might be retained by the algae than is excreted (Muscatine and Lenhoff 1965).

The data in Table 2 merely suggest that the ^{35}S-amino acids equilibrate with the unlabeled amino acids in the pool; the data do not suggest that the entire TCA-sol pool is greater than the alc-sol and alc-insol components. Since the gut of the anemone was not flushed out, much of this activity is probably unassimilated TCA-sol material. However, the presence of alc-insol, TCA-insol material in both the animal and algal fractions indicates that both the animal tissue and the zooxanthellae have assimilated free ^{35}S-amino acids.

Because the tone of this discussion is more about experiments that should follow rather than about any conclusions from those already performed, it is apparent that research on the nutritional contribution of the animal to the algal endosymbionts should be rewarding and full of surprises. Future experiments do not have to be limited to ^{35}S material, but should include tissue labeled with ^{32}P, radioactive metallic ions, or any isotope other than an isotope of carbon.

LITERATURE CITED

Gohar, H. A. F. 1940. Studies on the Xeniidae of the Red Sea. *Publications, Marine Biological Station, Ghardaqua* **2**: 25–118.

Hastings, J. W., L. Astrachan, and B. M. Sweeney. 1961. A persistent daily rhythm in photosynthesis. *J. General Physiology* **45**: 69–76.

Lenhoff, H. M. 1961. Digestion of ingested protein by *Hydra* as studied by radioautography and fractionation by differential solubilities. *Experimental Cell Research* **23**: 335–353.

McLaughlin, J. J. A., and P. A. Zahl. 1959. Axenic zooxanthellae from various invertebrate hosts. *Annals, New York Academy of Sciences* **77**: 55–72.

Muscatine, L., and H. M. Lenhoff. 1965. Symbiosis of hydra and algae. II. Effects of limited food and starvation on growth of symbiotic and aposymbiotic hydra *Biological Bulletin, Woods Hole* **129**: 316–328.

PART 4

CALCIFICATION

LEONARD MUSCATINE
University of California at Los Angeles

CHAPTER 22

Calcification in Corals

Calcification is widespread among the Coelenterata. Representatives of many orders secrete calcareous supportive elements (Table 1). Some medusae and certain hydroids secrete calcareous statoliths (Hyman 1940; Swedmark 1964). Experimental analyses of coelenterate calcification have been restricted largely to the true stony corals (order Scleractinia) and the hydrocorals (order Milleporina). This review briefly summarizes the major historical and current experimental research trends in coelenterate calcification, primarily in scleractinian corals.*

GENERAL MORPHOLOGY OF CORAL TISSUES IN RELATION TO THE SKELETON

Conventional techniques of microscopy and histology of corals (Hyman 1940) show four epithelial layers. Two layers make up the oral body wall which consists of an outer layer of secretory cells possessing distal microvilli and flagella (Goreau and Philpott 1956). Beneath it and separated by a fibrous mesoglea is a gastrodermal layer whose most conspicuous feature is the inclusion of symbiotic algae (zooxanthellae). Continuous with the oral body wall, but functionally distinct from it, is the aboral body wall consisting of gastrodermis, mesoglea, and epidermis. This epidermis, or calicoblastic epithelium, is the focal point of skeletogenesis.

SITE OF CALCIUM CARBONATE DEPOSITION IN CORALS

Heider in 1881 (see Ogilvie 1896) first described the lime-forming cells in corals to which he gave the name "calicoblast" cells. He thought the cells were of mesodermal origin, and claimed that the skeleton was initiated by intracellular precipitation of calcium carbonate, and that the

---

*I thank H. Lowenstam and S. A. Wainwright for advice on the preparation of this paper.

TABLE 1 Distribution of supportive skeletal types among coelenterates

Class	Order*	Form of Calcareous Skeleton	Mineral Composition
Hydrozoa	Stylasterina	Arborescent or encrusting; highly compact	Aragonite or Calcite
	Milleporina	As above	Aragonite
Anthozoa	Stolonifera	Loose or fused spicules	Calcite
	Telestacea	Spicules; organization diverse	Calcite
	Alcyonacea	Loose spicules	Calcite
	Coenothecalia	Massive; lamellae of large crystal "fibers"	Aragonite
	Gorgonacea	Spicules; axial skeleton in some species	Calcite, Aragonite
	Pennatulacea	Spicules; axial skeleton in some species	Calcite
	Madreporaria	Massive, compact exoskeleton	Aragonite

SOURCE: Hyman 1940; Revelle and Fairbridge 1957; Lowenstam 1954, 1964.

*Excludes those orders secreting statoliths only.

calicoblast itself transformed into a small body of calcareous fibers. In the same year, 1881, Koch (see Ogilvie 1896) postulated that ectodermal cells of coral secreted calcareous elements extracellularly. Koch's interpretations were confirmed and extended by Bourne (1887) and Fowler (1887, 1888; all in Ogilvie). They verified the existence of calicoblast cells in the ectoderm, noting further that only the aboral ectoderm secreted calcium carbonate.

Ogilvie (1896) vigorously upheld Heider's view that calicoblasts transform into bundles of calcareous fibers because the calcareous "scales" she observed on the surface of the skeleton had the same dimensions as intact calicoblasts. Hence, in her opinion, successive layers of calicoblasts were shed and replaced by new layers.

Hayashi (1937) described the calicoblast in corals as a single layer of cells with oval nuclei, and also described an amorphous organic membrane between the calicoblastic layer and the skeleton. In order to have a more favorable preparation for histological studies, Hayashi attempted to replace calcium carbonate crystals in the calicoblast with calcium oxalate. Yet his observations on the site of mineral deposition were inconclusive because he did not consider that calcium oxalate might be precipitated for reasons other than the replacement of calcium carbonate. Bryan and Hill (1941) maintained that the skeleton was an external secretion of the ectoderm. They postulated the site of mineralization to be a colloidal gel matrix secreted by the ectoderm, in which centers of calcification were initiated.

The site of precipitation of calcium carbonate in corals is still unknown. As yet there are no published histochemical or ultrastructural studies on the skeletogenic epithelium. It will be important to learn whether calcium precipitates intra- or extra-cellularly, and whether the calicoblastic epidermis has novel properties, discernible by microscopy, which may be related to calcification.

MICROMORPHOLOGY OF THE CORAL SKELETON*

Mineral Phase

Studies to date have not revealed any precipitated mineral constituents in the planula larva of corals during its pelagic stage. After attachment, the planula's aboral epithelium transforms from a tall glandular epithelium to a low squamous calicoblastic epithelium which deposits a thin ($1-2\mu$) plate of calcium carbonate interseptally. As the polyp develops, mineral is laid down in septal areas to form the six primary scleroseptae of the larval skeleton (Abe 1937; Wainwright 1963; see also Reed, this volume, Chapter 6).

*For a glossary of terms, see J. Wells, 1956, "Scleractinia," in *Treatise of Paleontology,* R. Moore, ed., vol. F, pp. 329 *et seq.*

In 1857 Milne-Edwards and Haime (see Ogilvie 1896) first described the micromorphology of the coral skeleton as a coalesced series of "sclerenchymatous nodules" that form trabeculae, rows of which give rise to lamellate structures. Ogilvie (1896) using the light microscope, described centers of calcification, their divergent crystal fibers, and the formation of trabeculae (see definition below). She stated: "Thus one might any day erect an exact model of the delicate calcareous framework of *Galaxea* using as the fundamental unit of structure a minute bush of fibres representing the last most highly calcified stage of the calicoblast." Ogilvie's "minute bush" is probably what Bryan and Hill (1941) refer to as a sperulite, "a radiating . . . aggregation of minerals, in an outward form approximating a spheroid and [resulting from] radial growth of prismatic or acicular crystals . . . about a common centre of inclusion." These authors regard spherulitic crystallization as the primary process of skeletal growth in the hexacorals.

Centers of calcification are defined as points of initiation of skeletal growth around which peripheral structures are formed. The exact nature of these centers with their crystalline units is not completely understood. Results of quantitative X-ray absorption studies show that there is no difference between the amount of calicum per unit volume in the calcification center and the amount in the peripheral skeleton. However, results of polarized light microscopy and X-ray microdiffraction on coral skeletons indicate that the aragonite in the center of calcification is in the form of randomly oriented submicroscopic crystals. From the center of calcification outwards there is a gradual tendency toward increased crystal size, increased preference of orientation, and aggregation into fibers (Wainwright 1964). The microscopically visible crystalline units of peripheral skeleton are polycrystalline aggregates which, because of their texture and the high degree of preferred orientation of their submicroscopic subunits, deserve to be called crystal fibers. X-ray microdiffraction shows that the crystals comprising the fibers deviate less than 10° from the fiber axis. The fibers have an irregular rather than a polyhedral outline in transverse section.

An aggregation of fibers arranged about an axis is defined as a trabecula. Such aggregations show a radial array of fibers in transverse section and divergent array in longitudinal section. A vertical aggregation of trabeculae gives rise to the various vertical elements of the skeleton, e.g., septa, pali, and columella. Horizontal elements such as tabulae and dissepiments are formed from parallel or "pilose" fibers at right angles to the horizontal surface (cf. Bryan and Hill 1941).

Analysis by X-ray microradiography reveals subtle quantitative differences in composition of regions of the coral skeleton. For example, the skeleton of the calyx in *Pocillopora damicornis* is slightly more dense than the coenosteal skeleton. This may reflect differences in the percentage or composition of organic matter associated with the skeleton in those areas. Few differences in skeletal mineralogy of hermatypic and ahermatypic corals (*Pocillopora damicornis* vs. *Lophelia obtusa,* respectively) have been noted (Wainwright 1964) despite the association of the former with symbiotic algae.

If the skeleton is deposited external to the living tissue, how is the species-specific architecture imposed? Bryan and Hill (1941) postulated that calcareous secretions are in contact with the secretory epithelium such that specific patterns are imposed by "organic guidance of an inorganic process."

Organic Phase

In the majority of those bony or calcareous skeletons which have been analyzed, the coral skeleton is intimately associated with an intrinsic intraskeletal organic matrix. Ogilvie (1896) referred to the organic component of the coral skeleton as the remains of calicoblast cells which had calcified and become incorporated into the skeleton. Duerden (1904; in Ogilvie 1907) noted an organic secretion external to the coral skeletogenic tissues, and implied that the skeleton arose externally by crystallization in this organic "matrix." Krempf (1907; in Ogilvie 1907) described the material remaining after decalcification as a "substratum of organic matter in the midst of which the calcareous part . . . has taken solid . . . the whole of the delicate organic meshwork forms a light translucent mass, similar to jelly and of extreme fragility."

Recently, Wainwright (1962, 1963) analyzed and characterized the matrix of the coral *Pocillopora damicornis.* After demineralization, an insoluble organic material remains, similar in microscopic detail to the intact skeleton, and constituting 0.01–0.1 percent of the dry weight of the skeleton. Its constituents are microscopically distinguishable as "(1) occasional filaments of lime-boring algae; (2) a loose, dispersed network of fibers about 1 μ in diameter; and (3) a transparent matrix that constitutes the bulk of the organic component." The latter consisted of an amorphous mass of smaller fibrils. Histochemical and microanalytical analyses, including X-ray diffraction, led to the conclusion that the matrix was chitin. This is the first record of this substance in the Anthozoa. The matrix was not shown to contain any soluble or fibrous protein, usually found covalently linked with chitin. When viewed by electron microscopy, the matrix appeared as a "spongework of fibrils," 20mμ in diameter, which were randomly oriented in a plane parallel to the skeletogenic epithelium and to the long axis of the skeletal branch. The precise spatial and functional relationships between matrix secretion and skeletal growth is not yet known.

Wainwright (1963) has put forth a number of challenging hypotheses concerning the role of the organic matrix in calcification. He suggested that the matrix may promote seeding or nucleation; chelate calcium; or control crystal size, shape, and orientation. He further postulated that the glucose moiety of the chitin in the matrix may be derived from photosynthetic products of the symbiotic algae associated with the corals. Lastly, he suggested that the limiting factor in calcification may well be the rate of synthesis of the matrix. In some recent experiments Muscatine (1967) isolated and identified the major photosynthetic product liberated by algae symbiotic with *Pocillopora damicornis* as glycerol. A small percentage of the soluble extracellular materials was provisionally identified as glucose. Further, studies by Pearse (this volume) and Muscatine and Cernichiari (1969) showed that the matrix of *Pocillopora* becomes labeled when the intact coral is incubated with $^{14}CO_2$. Radioactivity in the matrix is greater if corals are incubated in the light rather than in the dark, suggesting that, in the light, photosynthetic products of the symbionts are indeed incorporated into the matrix.

The relationship, if any, between the organic matrix as defined by Wainwright (1963) and the organic "membrane" figured by Hayashi (1937) immediately beneath the skeletogenic epithelium of *Pocillopora damicornis* and *Seriatopora caliendrum* is obscure. Goreau and Philpott (1956) also described a thin lamella external to the skeletogenic epithelium which stains like acid mucopolysaccharides. He suggested that this lamella acts both as an ion exchanger which absorbs calcium ions and as a "template" for calcium carbonate deposition (Goreau and Goreau 1959a). The views of Bryan and Hill (1941) that calcification may originate in an "ectodermal mucoid" or "gel precursor" are consistent with the existence of an extraepithelial membrane, but Bryan and Hill admit that they have not seen such an entity. Clearly, there is need for critical studies and operational definitions of matrix, membrane, and/or external colloidal gel. Such studies will have to deal with the histology of intact corals including their skeleton. Removal of tissues in preparation for a study of the skeleton may be expected to destroy any external membrane, and analysis of an isolated, skeleton-free coenosarc may not include this external membrane.

The existence of other organic components of the coral skeleton has been known for more than a century. Silliman (1846; in Wainwright 1963) recovered an ether-soluble wax from the skeleton of tropical reef corals. Bergmann and Lester (1940) obtained cetyl palmitate, sterols, cetyl alcohol, ketones, and low-melting-point hydrocarbons from the skeletons of *Acropora cervicornis* and *Manicina areolata.*

Coral matrix constituents are difficult to analyze chemically because of their low concentrations and because of the possibility of their being contaminated with tissues and lime-boring algae. Recent chemical analyses of matrices from five coral species are described by Young in this volume.

PHYSIOLOGY OF CALCIFICATION IN CORALS

A major breakthrough in studies on the physiology of calcification came when Goreau and Goreau (1959a) developed a rapid and precise technique for measuring calcium deposition in corals. Using ^{45}Ca they measured growth rates in various parts of coral colonies and observed rates of deposition under varied environmental conditions (see also Goreau and Goreau 1959b, 1960a, b).

Pathway of Calcium

Corals take calcium from seawater and deposit it in their skeleton (Goreau and Goreau 1959a). The effect of feeding on calcium deposition has not yet been investigated extensively. For example, food may influence calcification by stimulating organic growth, or by providing substrates which, when metabolized, increase the supply of carbonate ions. Most experiments on coral calcification have been carried out on nonfeeding corals.

Once ^{45}Ca is deposited in the skeleton of a living coral, it remains there permanently while the tissue is alive (Goreau and Goreau 1960b). If, however, the tissue is killed or removed, ^{45}Ca in the skeleton quickly exchanges at relatively high rates with environmental calcium. This rate of exchange is controlled by temperature (Goreau and Goreau 1960b).

Pathway of Carbonate

Studies using ^{14}C provide indirect evidence that carbonate in the skeleton may originate from both exogenous and endogenous CO_2. Goreau (1961) incubated corals with ^{45}Ca and ^{14}C-bicarbonate simultaneously. The labeled material deposited as skeleton had $^{45}Ca:^{14}C$ ratios higher than the theoretical stoichiometric mass ratio of 3.335. This irregular labeling pattern was interpreted as the result of the short incubation times which precluded the attainment of isotopic equilibrium in the system. The rate of deposition of ^{45}Ca was relatively higher because it underwent less isotopic dilution, and the rate for ^{14}C-bicarbonate lower because it experienced greater isotopic dilution. Goreau calculated that the endogenous pool of bicarbonate available for exchange with exogenous bicarbonate was 2–15 times larger than the exchangeable pool of calcium. This interpretation is vulnerable to the criticism that, in the light, photosynthesis by the symbiotic zooxanthellae will fix $^{14}CO_2$ and decrease the amount of $^{14}CO_2$ available for calcification.

$^{18}O:^{16}O$ and $^{13}C:^{12}C$ ratios in coral skeletal carbonate differ from those in seawater. Emiliani (1955) suggested that this difference may occur if oxygen produced by zooxanthellae during photosynthesis exchanges chemically with oxygen in the bicarbonate prior to incorporation of this bicarbonate in the skeleton. Photosynthesis may also affect fractionation of carbon isotopes (Park and Epstein 1960). Direct experimental evidence for incorporation of coral metabolic CO_2 into coral skeletal carbonate is given by Pearse in this volume. Although both endogenous and exogenous CO_2 may be deposited as carbonate, their relative quantitative importance and the conditions affecting their differential deposition are not known.

MINERALOGY OF THE CORAL SKELETON AND ENVIRONMENTAL FACTORS INFLUENCING CRYSTAL FORMATION

Stony corals deposit only aragonite, the orthorhombic metastable form of calcium carbonate, as do some other coelenterates, some marine calcareous algae, and some species of amphineuran, scaphopod, and adult cephalopod molluscs. Lowenstam (1954, 1964) suggested that temperature affects mineralogic composition of calcium carbonate produced by living organisms; and, in fact, aragonitic skeletons are found principally in warm-water organisms. The sparse growth and limited distribution of hermatypic corals in cold waters may result in part from inability to deposit sufficient aragonite. The relationship between temperature and calcification rate in the reef coral *Pocillopora damicornis* is described by Clausen in this volume.

According to Revelle and Fairbridge (1957), precipitation of calcium carbonate is favored by "(1) increase of temperature, which lowers the solubility product of calcium carbonate and also the solubility of carbon dioxide, thus increasing the carbonate ion concentration; (2) evaporation, which increases the calcium concentration and the carbonate alkalinity; (3) movement of supersaturated water into a region where nuclei or catalyzers of precipitation are present; (4) photosynthesis, which lowers the carbon dioxide content and hence increases the carbonate concentration; (5) bacterial production of ammonia or other weak bases, tending to raise the pH and hence to increase the carbonate concentration; (6) various, largely unknown, processes taking place in the body fluids or tissues of organisms, tending to increase the calcium or carbonate concentrations, to reduce the ionic strength, to produce nuclei, or otherwise catalyze carbonate precipitation." All of these processes may assist intrinsic calcification in hermatypic corals.

Quantitative data on factors influencing the intrinsic rate of calcification in reef corals have been marshalled by Goreau and Goreau (1959a, b). They found that light, the presence of symbiotic algae, and a specific inhibitor of carbonic anhydrase all significantly influence calcification rates in corals. A working hypothesis was based on the following reaction:

$$Ca^{++} + 2HCO_3^- \rightarrow Ca(HCO_3)_2 \rightarrow CaCO_3 + H_2CO_3$$

The reaction velocity may be increased by removing the end product, carbonic acid, whose conversion to water and carbon dioxide is catalyzed by the enzyme carbonic anhydrase. In the light, with algae present, the carbon dioxide is consumed by photosynthesis. A carbonic anhydrase inhibitor "decelerates" calcification by obstructing the removal of the carbonic acid end product.

Another interpretation for the role of algae is offered by Simkiss (1964a, b) based upon the interesting observation that, when corals with algae and corals without algae are both placed in the dark, the corals with algae still calcify significantly faster than those without. He suggests that orthophosphate and organic phosphates, such as glycerophosphate, act as crystal poisons and therefor interfere with the formation of aragonite crystals. Simkiss believes that the symbiotic algae in corals remove phosphates from the microenvironment, even in darkness, as a normal concomitant of their nutritional activities, thus creating a more favorable environment for calcification. He suggests also that corals without algae, in darkness or light, have the lowest calcification rates, presumably, in part, because phosphates accumulate and deter crystallization.

Rates of calcification vary in different parts of a particular colony. Invariably, especially in branching species, growing tips calcify much faster than proximal portions (Goreau and Goreau 1959a). Apical polyps in some acroporids calcify faster than lateral ones, even though the apical ones have far fewer zooxanthellae than the latter (Goreau 1963). This observation suggests that in acroporid tips the intrinsic rate of calcification is high.

Regular periodic variations in the rates of calcium deposition are inferred from the occurrence of bands on the epitheca of many modern and fossil corals. These bands may occur as coarse ridges, thought to represent annual or monthly increments, and as fine lines, up to 60 per millimeter, which may represent daily growth increments. The number of "daily" bands between each "annual" band in fossil vs. modern corals has been used to count the number of days in the year. For example, Devonian corals average 400 daily bands between each annual band compared to 360 in modern corals (Wells 1963; Scrutton 1964), which leads to the inference that the Devonian day was shorter than the present day. This interpretation is consistent with the geophysical theory that day length has been changing by about 2 seconds every 100,000 years due to slowing down of the earth's rate of rotation as a result of tidal friction. What aspects of coral biology give rise to growth with such increments? Goreau has suggested that the periodicity of feeding and the effect of zooxanthellae may be the underlying stimuli. His interpretation, however, does not account for the appearance of growth increments in deep dwelling, ahermatypic corals. An intrinsic biological-clock mechanism controlling rate of calcification has also been suggested (Stubbs 1966).

Inasmuch as zooxanthellae play an important role in calcification of reef corals, it follows that the calcification rate of a given coral should be influenced by its depth in the ocean. Goreau (1963:162) stated that "one of the most interesting problems of environmental control of reef coral growth is the effect of ambient light intensity on colony mass and form. Corals growing as . . . large hemispheroidal heads in shallow water become progressively flatter and less massive as the depth increases. . . . In deep water, natural selection seems to favor coral with thin, light skeletons."

Ecological studies show that reduction in light intensity is a primary factor controlling change in colony shape. Corals in dimly lit areas are fragile and are not good reef-builders. No reef-building occurs below 46 meters (Wells 1957). Goreau (1963) suggested that, although calcification decreases as the amount of light decreases, tissue growth is sustained at normal levels with an adequate food supply. As a result of low light intensity, the coral forms a flattened skeleton with a larger polyp surface area relative to skeletal mass. An important area for future investigation will be a quantitative study of the role of zooxanthellae in coral organic productivity as influenced by depth.

SKELETAL CALCIFICATION IN OTHER COELENTERATES

Kawaguti (1964) has recently investigated by electron microscopy the morphology and growth of the spicules in certain alcyonarians. Scleroblasts in the ectoderm and endoderm contain small vesicles about 1μ in diameter in which calcareous elements are thought to be formed. The presumed precursors of the calcified elements in the scleroblasts appear first as aggregations of threads about 50 angstroms in diameter. As the scleroblasts increase in size, several of them coalesce to yield the mature spicule which ultimately appears in the mesoglea. Little else is known of the physiology of calcification in these organisms.

LITERATURE CITED

Abe, N. 1937. Post-larval development of the coral *Fungia actiniformis* var. *palawensis* Döderlein. *Palao Tropical Biological Station Studies* **1**: 73–93.
Bergmann, W., and D. Lester. 1940. Coral-reefs and the formation of petroleum. *Science* **99**: 452–453.
Bryan, W. H., and D. Hill. 1941. Spherulitic crystallization as a mechanism of skeletal growth in the hexacorals. *Proceedings, Royal Society of Queensland* **52**: 78.
Emiliani, C. 1955. Pleistocene temperatures. *J. Geology* **63**: 538–578.
Goreau, T. F. 1961. On the relation of calcification to primary production in reef building organism. In *The biology of hydra and of some other coelenterates: 1961,* H. M. Lenhoff and W. F. Loomis, eds., pp. 269–285. Coral Gables: University of Miami Press.
———. 1963. Calcium carbonate deposition by coralline algae and corals in relation to

their roles as reef-builders. *Annals, New York Academy of Sciences* **109**: 127–167.

Goreau, T. F., and N. I. Goreau. 1959a. The physiology of skeleton formation in corals. I. A method for measuring the rate of calcium deposition by corals under different conditions. *Biological Bulletin, Woods Hole* **116**: 59–75.

———. 1959b. The physiology of skeleton formation in corals. II. Calcium deposition by hermatypic corals under various conditions in the reef. *Biological Bulletin, Woods Hole* **117**: 239–250.

———. 1960a. The physiology of skelton formation in corals. III. Calcification rate as a function of colony weight and total nitrogen content in the reef coral *Manicina areolata* (Linnaeus). *Biological Bulletin, Woods Hole* **118**: 419–429.

———. 1960b. The physiology of skeleton formation in corals. IV. On isotopic equilibrium exchanges of calcium between corallum and environment in living and dead reef-building corals. *Biological Bulletin, Woods Hole* **119**: 416–427.

Goreau, T. F., and D. E. Philpott. 1956. Electron micrographic study of flagellated epithelia in madreporarian corals. *Experimental Cell Research* **10**: 552–556.

Hayashi, K. 1937. On the detection of calcium in the calicoblasts of some reef corals. *Palao Tropical Biological Station Studies* **1**: 169–176.

Hyman, L. H. 1940. *The invertebrates.* I. *Protozoa through Ctenophora.* New York: Academic Press.

Kawaguti, S. 1964. Electron microscopy on the spicules and the polyp of a gorgonian, *Euplexaura erecta. Biological J. of Okayama University* **10**: 23–38.

Lowenstam, H. 1954. Factors affecting the aragonite:calcite ratios in carbonate-secreting marine organisms. *J. Geology* **62**: 284–322.

———. 1964. Coexisting calcites and aragonites from skeletal carbonates of marine organisms and their strontium and magnesium contents. In *Recent researches in the fields of hydrosphere atmosphere and nuclear geochemistry,* Miyake and Koyama, eds. Tokyo: Maruzen Press.

Muscatine, L. 1967. Glycerol excretion by symbiotic algae from corals and the *Tridacna* and its control by the host. *Science* **156**: 516–519.

Muscatine, L., and E. Cernichiari. 1969. Assimilation of photosynthetic products of zooxanthellae by a reef coral. *Biological Bulletin, Woods Hole* **137**: 506–523.

Ogilvie, M. 1896. Microscopic and systematic study of madreporarian types of corals. *Philosophical Trans., Royal Society. London,* Ser. B, **187**: 83–345.

———. 1907. Note on the formation of the skeleton in the Madreporaria. *Quarterly J. Microscopical Science* **51**: 473–482.

Park, R., and S. Epstein. 1960. Carbon isotope fractionation during photosynthesis. *Geochimica et Cosmochimica Acta* **21**: 110–126.

Revelle, R., and R. Fairbridge. 1957. Carbonates and carbon dioxide. *Geological Society of America* **67**: 239–295.

Scrutton, C. T. 1964. Periodicity in Devonian coral growth. *Paleontology* **7**: 552–558.

Simkiss, K. 1964a. Possible effects of zooxanthellae on coral growth. *Experientia* **20**: 140.

———. 1964b. The inhibitory effects of some metabolites on the precipitation of calcium carbonate from artificial and natural sea water. *J. du Conseil* **29**: 6–18.

Stubbs, P. 1966. Coral timekeepers of the slowing earth. *New Scientist* **29**: 828–829.

Swedmark, B. 1964. The interstitial fauna of marine sand. *Biological Reviews, Cambridge Philosophical Society* **39**: 1–42.
Wainwright, S. A. 1962. An anthozoan chitin. *Experientia* **18**: 18.
——. 1963. Skeletal organization in the coral *Pocillopora damicornis. Quarterly J. Microscopical Science* **104**: 169–183.
——. 1964. Studies of the mineral phase of coral skeleton. *Experimental Cell Research* **32**: 213–230.
Wells, J. W. 1957. Coral reefs. *Geological Society of America.* Memoirs, no. 67, pp. 609–632.
——. 1963. Coral growth and geochronometry. *Nature* **197**: 948–950.

VICKI BUCHSBAUM PEARSE CHAPTER 23
Stanford University, Stanford, California

Sources of Carbon in the Skeleton of the Coral *Fungia scutaria*

Two major possible sources for the carbonate in the calcium carbonate contained in coral skeletons are soluble carbonates from seawater and carbon dioxide produced by the metabolism of the coral tissues. Whereas radioisotopic tracer experiments have established that carbonate from seawater is incorporated into the skeleton by many corals (Goreau 1961, 1963), evidence for metabolic CO_2 being incorporated into the skeleton is much less direct. Goreau (1961) found that calcification rates measured by the incorporation of radioactive carbonate from seawater are usually lower than those measured by using radioactive calcium; he proposed that the labeled carbonate may be diluted by unlabeled carbonate in the coral tissues. In addition, findings that the ratios of ^{13}C to ^{12}C and of ^{18}O to ^{16}O in seawater differ from those in the coral skeleton suggest that not all the skeletal carbonate originates from carbonate in seawater (Craig 1953; Emiliani 1955; Keith and Weber 1965; Lowenstam and Epstein 1957).

In order to examine further the possibility that some skeletal carbonate may originate from metabolic CO_2, I fed ^{14}C-labeled mouse tissue to small individuals of the coral *Fungia scutaria,* a solitary polyp containing abundant symbiotic algae. My experiments offer direct evidence that metabolic $^{14}CO_2$ is incorporated into skeletal carbonate in this coral. I also present data in this paper on the effects of light and starvation on calcification, and information concerning the origin of the coral matrix.

MATERIALS AND METHODS

Solutions of $Na_2{}^{14}CO_3$ and $^{45}CaCl_2$ (New England Nuclear Company) were added to freshly filtered seawater so that 40-ml portions, in 50-ml glass jars, contained approximately 25 µc $Na_2{}^{14}CO_3$ and 40 µc $^{45}CaCl_2$ per jar. For experiments in which the corals were to be kept in the light, the jars were covered only with a transparent petri dish. For experiments in which the corals were to be kept in the dark, the jars were wrapped with black electrical tape and were covered with black caps and aluminum foil.

Specimens of small, stalked *Fungia scutaria* (1-2 cm diameter, 300-1,000 mg dry weight) were placed in each jar. The jars were kept at 28° C ± 0.5° C by a bath of circulating tap water, set above a bank of four 40-watt Sylvania Lifeline fluorescent tubes. The incident illumination was approximately 2,000 footcandles. After 24 hours' incubation, the corals were transferred to clean seawater and analyzed.

Labeling Corals by Feeding Them Radioactive Mouse Tissues

A laboratory mouse was injected intraperitoneally with 0.5 mc of ^{14}C-labeled protein hydrolysate in 0.1 ml physiological saline, and was killed 24 hours later. Pieces of liver, kidney, and intestine were fed to small *Fungia* (90-240 mm diameter, 150-1,650 mg dry weight). The animals actively swallowed and retained the mouse tissue if seawater suspensions of crushed *Artemia* nauplii were offered at the same time. Residual fragments of mouse tissue were sometimes observed to be egested by the corals several hours after they were fed. Mouse tissue offered to the corals without the *Artemia* suspension was seldom accepted and was never retained for more than a few minutes. Feeding was discontinued for several days (see Table 1) before the specimens were analyzed.

Analysis of Specimens

Each specimen was placed in a small beaker and just covered with concentrated (58%) ammonium hydroxide. The beakers were heated in a water bath at 60°-70° C for about 1 hour. Most of the loosened tissue readily came off the skeleton when repeatedly washed with jets of ammonium hydroxide solution. The skeleton was taken out, washed under a strong jet of tap water to remove remaining fragments of tissue, dried in an oven at 100° C, and weighed.

The tissue in ammonium hydroxide was homogenized with a Teflon-coated pestle and brought to prescribed volume. Portions were removed, placed on planchets, acidified with HCl, and counted on a Nuclear-Chicago gas flow detector, model 470. Samples were also taken for protein determination by a modified Lowry method.

The clean skeletons were dissolved in 6 N HCl in an evacuated flask, and the CO_2 evolved was recovered in a small beaker of 5 N KOH which was on the bottom of the flask (Doty and Oguri 1959). When all the carbonate was dissolved, an additional hour was allowed for the carbon dioxide to become more completely absorbed; the flask was opened and samples of the KOH solution were removed for counting. The acid-insoluble residue, which included the organic matrix of the skeleton, was collected on a Millipore filter, dried, and counted. The acid filtrate was brought to a known volume, and portions were removed for counting.

TABLE 1 Distribution of radioactivity in *Fungia scutaria* fed mouse tissue labeled with ^{14}C-amino acids

	Days Fasted after Labeling	Tissue		Skeletal Carbonate		Acid-insoluble Residue	
		Activity*	% Activity	Activity*	% Activity	Activity*	% Activity
Group 1 (3 individual corals)	4–6	5,374	95.1	245	4.3	34	0.6
Group 2 (4 individual corals)	13	4,683	90.0	485	9.3	35	0.7

*Activity measured as cpm ($\log_{10}$ mg skeletal weight X number of feedings).

Skeletal ^{45}Ca was determined from the acid filtrate. Control experiments using ^{45}Ca and ^{14}C label separately showed that the acid filtrate did not contain measurable amounts of ^{14}C. The small amount of ^{45}Ca in the tissue fraction affects ^{14}C data significantly only for experiments carried out in the dark, a condition under which ^{14}C incorporation is low. Therefore, ^{14}C values from experiments carried out in the dark were corrected for the presence of ^{45}Ca. About 95 percent of the ^{14}C from skeletal carbonate was recovered.

RESULTS

Incubation of Corals with Na$_2$^{14}CO$_3$ and ^{45}CaCl$_2$ in Seawater

Specimens of *Fungia* were incubated in the light or dark for 24 hours in seawater containing Na$_2$^{14}CO$_3$ and ^{45}CaCl$_2$. After the incubation, each coral was rinsed in seawater, and the tissue (with its contained symbiotic algae) was separated from the skeleton. The cleaned skeleton was analyzed as described under "Methods." The results are shown in Table 2.

Almost 40 times more label was incorporated into the tissue in the light than in the dark, presumably because of the photosynthetic activity of the algae. Also stimulated significantly (12-fold) by light was the amount of label in the acid-insoluble residue. Part of this increase may be due to contamination with fragments of labeled tissue, but the increase may also reflect labeling of the coral's organic matrix with radioactive material obtained from the photosynthetic products of the algae.

The calcification rate was 4–6 times greater in the light than in the dark, as measured by incorporation of both labeled carbonate and calcium. Control experiments showed that the increase in the ^{14}CO$_2$ released when the skeleton was dissolved in acid orginated from skeletal carbonate, and not from any product of algal photosynthesis that might have contaminated the skeleton.

TABLE 2 Distribution of radioactivity in *Fungia scutaria* labeled with Na$_2$^{14}CO$_3$ and ^{45}CaCl$_2$

	In Light	In Darkness	Ratio Light:Dark
Tissue ^{14}C	5,880	163	36:1
Skeletal Carbonate ^{14}C	993	250	4:0
Acid-insoluble Residue ^{14}C	12.1	1.0	12:1
Skeletal Calcium ^{45}Ca	1,110	182	6:1

NOTE: Radioactivity measured as cpm/mg skeletal weight. Three individual corals per experiment were used.

Labeling of Corals by Feeding Them Mouse Tissue Labeled with ^{14}C-Amino Acids

Pieces of mouse tissue labeled with ^{14}C-amino acids were fed to small *Fungia* over a period of days. The animals were then fasted: one group for 4–6 days, another for 13 days. At the end of the period of fasting, the tissue and skeletal components were separated as before. The data, presented in Table 1, are expressed either as cpm/(log$_{10}$ mg skeletal weight × number of feedings), or as the percentage of those activities in each fraction. The activities were corrected in this way because the animals were of different sizes and because it was not possible to feed each polyp with precisely the same amount of labeled food. The rationale for my corrections is as follows: (1) Skeletal weights were used as an index of size. Within the size range of animals used, skeletal weight and tissue nitrogen values were found to be linearly related in *Fungia,* as in *Pocillopora* (see Clausen, this volume). (2) Because all animals received similarly sized pieces of tissue at each feeding, correction of total activity by weight or nitrogen measurements gives low values for larger animals. Corrections by log weight yield more comparable values, and were therefore made. (3) As all animals did not receive the same number of feedings, a correction for this variable also yields more comparable values.

The results show that a significant portion of the activity was consistently found in the carbonate of the skeleton, offering direct evidence that metabolic CO_2 is incorporated into the carbonate (Table 1). In the corals fasted for 13 days (group 2), the amount of radioactivity in the skeletal carbonate was twice that of group 1, while the amount of radioactivity in the tissue decreased. Hence, the animals continued to calcify during this period although they were not being fed, and in doing so fixed some of their metabolic CO_2.

The data suggest that some of the ^{14}C derived from the food was incorporated into the organic matrix of the skeleton also. The activity in this fraction did not change significantly after the animals were fasted for longer periods.

DISCUSSION

The light/dark ratios obtained with *Fungia* for fixation of $^{14}CO_2$ in the tissues and for skeletal calcification are similar to those obtained by Goreau for other coral species. Thus, in *Fungia* too, light stimulates calcification, and presumably the symbiotic algae play some role in enhancing the calcification rate. My experiments (Table 1) provide direct evidence that metabolic CO_2 is deposited in the skeleton as carbonate. This finding supports Goreau's interpretation of his double-labeling experiments in which he found that calcification rates calculated from the ^{14}C in skeletal carbonate were lower than those calculated from skeletal

^{45}Ca. He interpreted the lower carbonate values as being the result of isotopic dilution of the exogenous carbonate pool with unlabeled CO_2 in the coral tissues. It is important to point out, however, that the relative contributions of carbonate from seawater and from metabolism are still unknown; the sources of CO_2 for mineral deposition at any one instant might be expected to vary with such factors as the degree of illumination and the nutritional state of the animal.

Additional experiments on the labeling of the acid-insoluble residue may help to elucidate whether the coral matrix is synthesized by the animal tissue independently of the algae, or whether the symbiotic algae provide the components of the matrix as proposed by Wainwright (1963). It will be necessary, however, to be certain that the matrix is not contaminated with coral tissue, and to identify the components of the matrix as has been done with other corals (Wainwright 1962; also Young, this volume). Once this has been determined, then experiments in which the corals are labeled either with $Na_2{}^{14}CO_3$ or with ^{14}C-labeled food, and kept either in the light or in the dark, should provide answers regarding the origin of substrates for the matrix. Since the matrix is considered necessary for mineral deposition, then knowledge of its composition and of conditions controlling its synthesis should add to our understanding of factors regulating calcification.

SUMMARY

1. Direct evidence for the incorporation of metabolically produced CO_2 into skeletal carbonate was obtained by feeding ^{14}C-labeled food to corals and determining the $^{14}CO_2$ released from their skeletons.

2. Light increased the amount of label from $Na_2{}^{14}CO_3$ and $^{45}CaCl_2$ incorporated into the skeleton of *Fungia scutaria* 4–6 times over that taken up in the dark.

3. Unfed corals continued to calcify for several days, depositing metabolically produced $^{14}CO_2$ into the skeleton.

4. The acid-insoluble organic matrix of the skeleton was found to be labeled with ^{14}C originating from exogenous $Na_2{}^{14}CO_3$. Carbon-14 from coral metabolism of ^{14}C-labeled food may also be incorporated into the skeletal organic matrix.

5. A method is described for fractionating *Fungia* into tissue, skeleton, and matrix.

LITERATURE CITED

Craig, H. 1953. The geochemistry of the stable carbon isotopes. *Geochimica et Cosmochimica Acta* **3**: 53-92.

Doty, M. S., and M. Oguri. 1959. The carbon-14 technique for determining primary plankton productivity. *Pubblicazioni della Stazione zoologica di Napoli* **31** suppl **70**: 94.

Emiliani, C. 1955. Pleistocene temperatures. *J. Geology* **63**: 538-578.

Goreau, T. F. 1961. On the relation of calcification to primary productivity in reef building organisms. In *The biology of hydra and of some other coelenterates: 1961*, H. M. Lenhoff and W. F. Loomis, eds., pp. 269-285, Coral Gables: University of Miami Press.

———. 1963. Calcium carbonate deposition by coralline algae and corals in relation to their roles as reef-builders. *Annals, New York Academy of Sciences* **109**: 127-167.

Keith, M. L., and J. N. Weber. 1965. Systematic relationships between carbon and oxygen isotopes in carbonates deposited by modern corals and algae. *Science* **150**: 498-501.

Lowenstam, H. A., and S. Epstein. 1957. On the origin of sedimentary aragonite needles of the Great Bahama Bank. *J. Geology* **65**: 364-375.

Wainwright, S. A. 1962. An anthozoan chitin. *Experientia* **18**: 18-19.

———. 1963. Skeletal organization in the coral, *Pocillopora damicornis*. *Quarterly J. Microscopical Science* **104**: 169-183.

CONRAD CLAUSEN CHAPTER 24
Loma Linda University, Loma Linda, California

Effects of Temperature on the Rate of ^{45}Calcium Uptake by *Pocillopora damicornis*

Observations on the distribution of hermatypic corals have indicated that temperature markedly affects the distribution and richness of reefs. Reef formation occurs only in relatively warm waters, being most luxuriant in the vicinity of the equator. To date no rigorous laboratory studies have been made on the effects of temperature on the rate of calcification in reef corals. I undertook to determine how the deposition of ^{45}Ca into ^{45}CaCO$_3$ by corals was affected by temperature.

MATERIALS AND METHODS

The organism used, *Pocillopora damicornis*, was obtained from Kaneohe Bay, Oahu, Hawaii in the summer of 1967. Corals having long slender branches from which fairly uniform branch tips could be obtained were collected. The colonies were maintained in running seawater tables from 1 to 8 days before use.

Quantification of Coral Tissue

Such methods as dry weight or wet weight could not be used as a measure of metabolically active tissue in coral tips because of the great mass contributed by the calcified skeleton. Although an organic nitrogen determination by the micro-Kjeldahl method as used by Goreau (1959) is perhaps the most satisfactory, other indirect methods entailing less time might be just as satisfactory under conditions involving a single species of coral with samples of similar size, shape, and position in the coral colony.

If dry weight of the skeleton or number of calices is directly proportional and closely correlated with the organic nitrogen of the tissue, these three measurements can be used interchangeably, or any one can be used in estimating the others. Figures 1, 2, and 3 show that such a correlation exists. To obtain these relationships 10 tips varying in skeletal weight from 8.7 mg to 257.4 mg and having various shapes (branched and unbranched) were removed from a colony of *Pocillopora damicornis*.

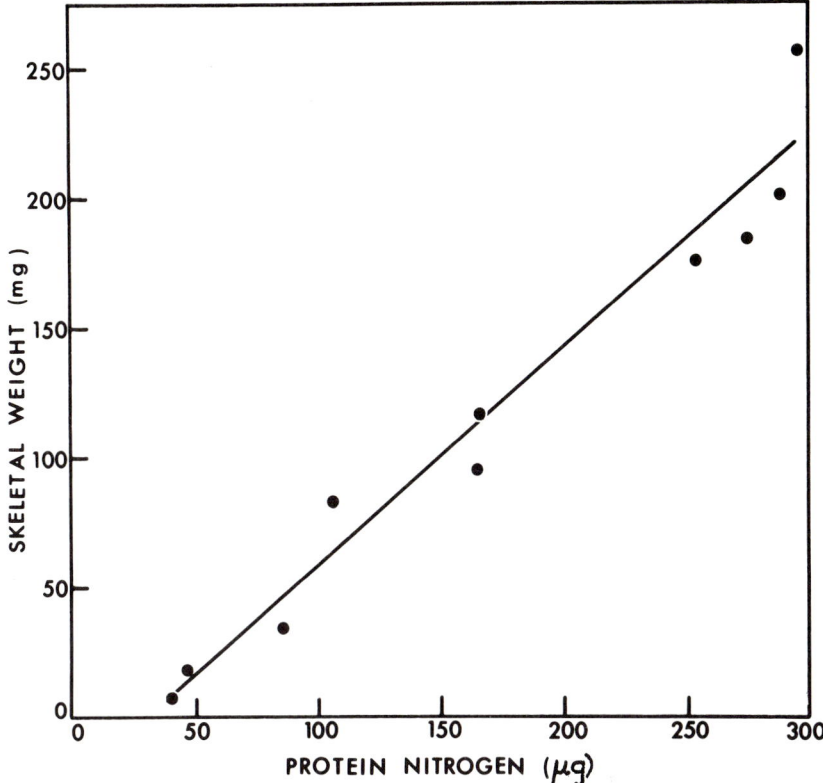

FIG. 1. Correlation between protein nitrogen and dry skeletal weight of the coral *Pocillopora damicornis*. Ten sample tips used for this graph are the same as those used for the graphs in Figures 2 and 3. The tips cover a dry skeletal weight range from 8.7 mg to 257.4 mg. Correlation coefficient is greater than 0.97.

Measure of Organic Nitrogen

The tissue nitrogen was solubilized by first freezing the piece of coral in 1.5 ml of distilled water and then, after thawing the coral, adding 1.5 ml of concentrated (14.8 M) NH_4OH and heating at 65° C for an hour. The test tubes were vibrated briefly to dislodge any remaining tissue adhering to the skeleton. The skeletal pieces were removed and rinsed with distilled water; the rinse solution was saved and added back to the NH_4OH digest. The combined solutions were diluted to 5 ml (or 10 ml for the larger tips), and 0.5-ml portions were removed for protein determination.

The washed skeletal pieces were dried. The dry weights were taken and the number of calices were counted.

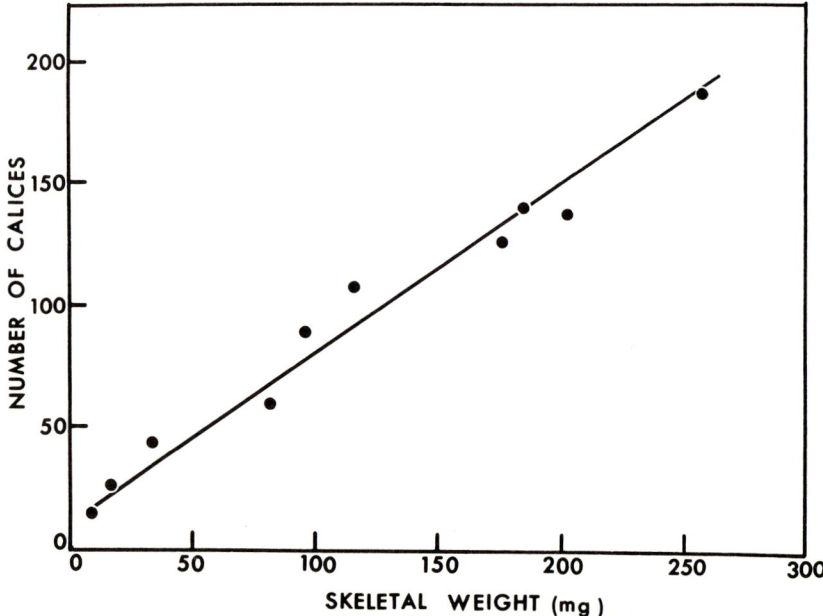

FIG. 2. Correlation between skeletal weight and numbers of calices of the coral *Pocillopora damicornis*. Ten sample tips used for this graph are the same as those used for the graphs in Figures 1 and 3. The tips cover a dry skeletal weight range from 8.7 mg to 257.4 mg. Correlation coefficient is greater than 0.97.

Preparation of Coral Branches for Incubation with ^{45}Ca

Coral branches having 5–10 tips that might be removed for assay of radioactivity were cut from a coral head and were placed upright in the experimental vessel by inserting the stem of the branch into a plastic stand. The branches were given a bountiful supply of nauplii of the brine shrimp *Artemia salina*, were left in the dark for 30 minutes, and were then returned to the seawater table for 30 minutes so that extraneous or regurgitated *Artemia* would be washed off. The plastic stand with the coral branch was then transferred to a glass jar containing several hundred milliliters of nonradioactive seawater. (The coral was kept submerged throughout all the above manipulations.)

To reduce any possible shock involved in the temperature changes for the experiment, the seawater containing the corals was brought slowly to the experimental temperature over a period of about 1 hour. Next, the coral branches were transferred to about 250 ml filtered seawater (at the experimental temperature) enriched with ^{45}CaCl$_2$ (New England Nuclear Company). Sufficient ^{45}CaCl$_2$ was used so that 50-μl portions of the labeled seawater gave approximately 25,000 cpm on a Nuclear Chicago gas flow counter, model 1042. The label was added to the seawater just prior

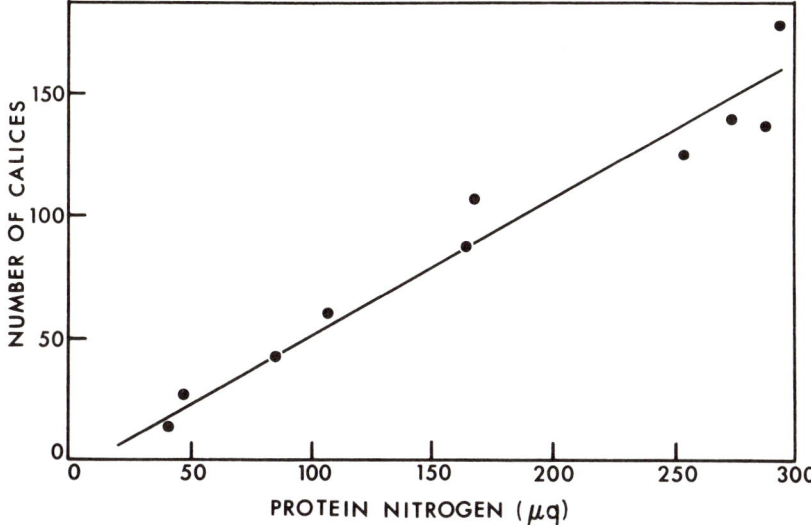

FIG. 3. Correlation between protein nitrogen and number of calices of the coral *Pocillopora damicornis*. Ten sample tips used for this graph are the same as those used for the graphs in Figures 1 and 2. The tips cover a dry skeletal weight range from 8.7 mg to 257.4 mg. Correlation coefficient is greater than 0.97.

to each experiment. To permit aeration and circulation, the glass jar serving as the incubation vessel was fitted with a two-hole rubber stopper; through a glass tube in one of the holes a constant stream of air was bubbled into the water.

The coral branches did not seem to suffer any immediate ill effects from being handled in this manner and from being detached from the rest of the colony. To the contrary, corals incubated at 20° C and 25° C (Table 1) appeared healthy for more than a week after an experiment.

Control of Temperature and Light

Water from a temperature-controlled bath was circulated through a Plexiglas box which contained the experimental jars. The main light source was four 40-watt fluorescent bulbs kept beneath the Plexiglas box and covered with a translucent plastic shield to give a diffused and even light. In addition there was ambient light coming in on the sides. The incident light intensity on the bottom of the jar was approximately 2,000 footcandles.

Sampling of the Coral Branch for Radioactivity

At the appropriate time intervals the sample tips were cut off the branches for assay. Most of the samples used weighed from 10–60 mg (dry skeletal weight). The samples were blotted briefly on filter paper to

TABLE 1 Effect of temperature on ^{45}Ca uptake by *Pocillopora damicornis*

Temperature of Incubation (Centigrade)	Incubation Time (hours)	Number of Samples	Days Maintained on Water Table before Experiment	^{45}Ca Uptake (cpm/mg of skeletal weight ± SE)	Days Survived after Experiment
12°	1.0	9	8	5.8 ± 0.6	<1
	3.0	10	8	8.2 ± 1.1	<1
	6.0	10	8	12.6 ± 1.6	<1
15°	3.0	7	5	20.5 ± 3.7	<2
	6.0	7	5	30.4 ± 7.2	<2
20°	3.0	7	4	89.8 ± 21.6	>8
	6.0	7	4	93.8 ± 20.3	>8
25°	0.5	10	4 (30 samples)	35.8 ± 7.8	>6
	1.0	10	6 (8 samples)	77.7 ± 17.5	>6
	1.5	10	1 (12 samples)	93.5 ± 23.7	>6
	3.0	10		219.5 ± 52.1	>6
	6.0	10		407.9 ± 84.1	>6
30°	3.0	10	7	127.4 ± 24.2	—
	6.0	9	7	193.0 ± 46.0	—

remove excess seawater and then rinsed five times for 2 minutes each in 2-ml portions of distilled water. They were put in 0.5 ml of distilled water and frozen until they were prepared for radioactive assay.

Measurement of ^{45}Calcium in Skeleton

The tissue was removed from the skeleton by adding 0.5 ml of concentrated (14.8 M) NH_4OH to the 0.5 ml distilled water already present, and heating the solution to 65° C for an hour. After heating, the skeleton was removed and rinsed six times in distilled water to free it of unincorporated ^{45}Ca. The samples were dried, weighed, and dissolved in 0.5 ml of 3 N HCl (1.0 ml was necessary for the larger samples), and 0.1-ml portions of the resultant solution were put on planchets. Each planchet was fitted with a disc of lens paper to insure even spreading of the solution over the planchet. The planchets were dried and counted.

RESULTS

Rate of ^{45}Ca Uptake at Different Temperatures

Table 1 summarizes the experiments in which the uptake of ^{45}Ca by *Pocillopora damicornis* at different temperatures was measured. Most of the data obtained was from samples taken at the end of 3- and 6-hour incubation periods. From this data the amount of ^{45}Ca incorporated per hour was calculated and these rates were plotted against temperature (Figure 4). Those rates based on samples taken after a 3-hour incubation period were consistently higher at all temperatures than those rates based on samples taken after 6 hours of incubation. There is apparently a decrease in the rate of ^{45}Ca incorporated with increasing incubation time. The patterns of the curves, however, are similar for both incubation periods. As seen in the figure, the rate at which ^{45}Ca was incorporated increased exponentially with temperature from 12° C to 25° C, whereas at 30° C it was again lower.

At 12° C and 25° C data were also obtained for earlier incubation times (Table 1). These data were combined with the 3- and 6-hour data in Figure 5. The rate at 25° C was about 50 times that at 12° C. (Note the different scale in Figure 5 for the 12° C experiment.) At 12° C, although the amount of ^{45}Ca incorporated increases linearly with time from 1 to 6 hours of incubation, the rate during this period is apparently less than that during the first hour. Again, at 25° C the amount of ^{45}Ca incorporated appears to increase linearly with time, but this may be more apparent than real since the first three points are based on a different experiment (using different coral colonies which were maintained in the laboratory for a different length of time—4 days) than the last two points. Most of the data indicate a definite decrease at all temperatures in the rate of ^{45}Ca incorporation with increasing incubation time.

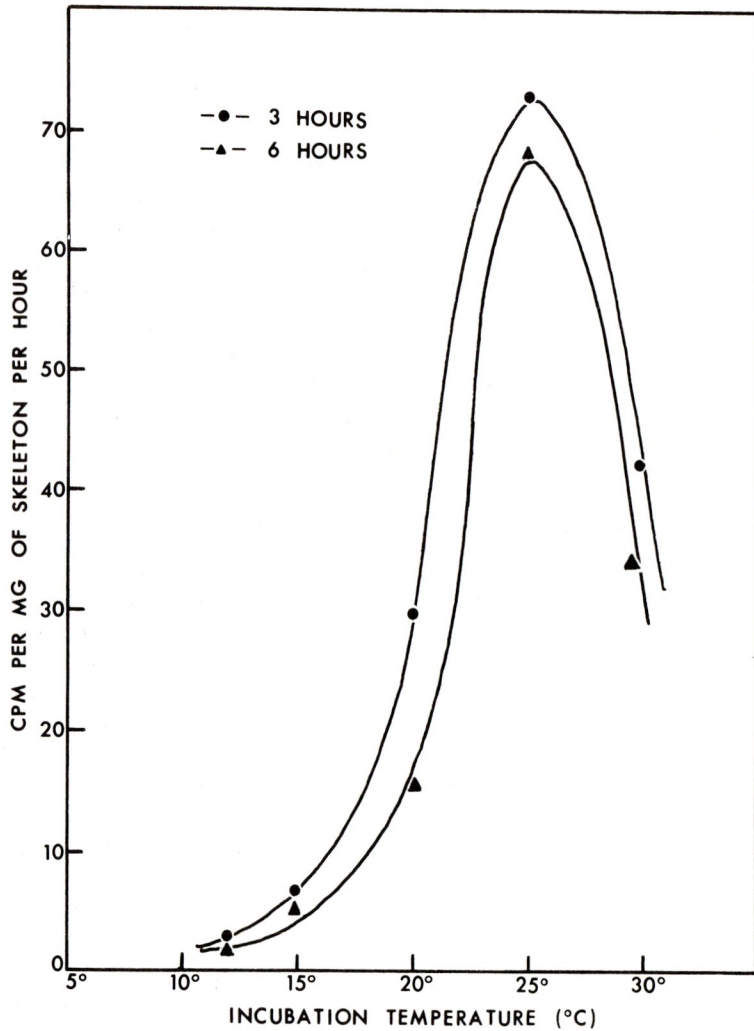

FIG. 4. The effect of temperature on the rate of incorporation of ^{45}Ca by *Pocillopora damicornis*. The lines are based on data from 3- and 6-hour incubation periods.

Arrhenius Plot of Data

To study the energy of activation an Arrhenius plot of the data was made. The log of the rate (or velocity) of ^{45}Ca incorporation was plotted against the reciprocal of the absolute temperature (Figure 6). Generally in such plots either the rate constant or initial velocity is used rather than the average velocity. It is assumed here, however, that the average velocity is proportional to both the initial velocity and the rate constant, inasmuch

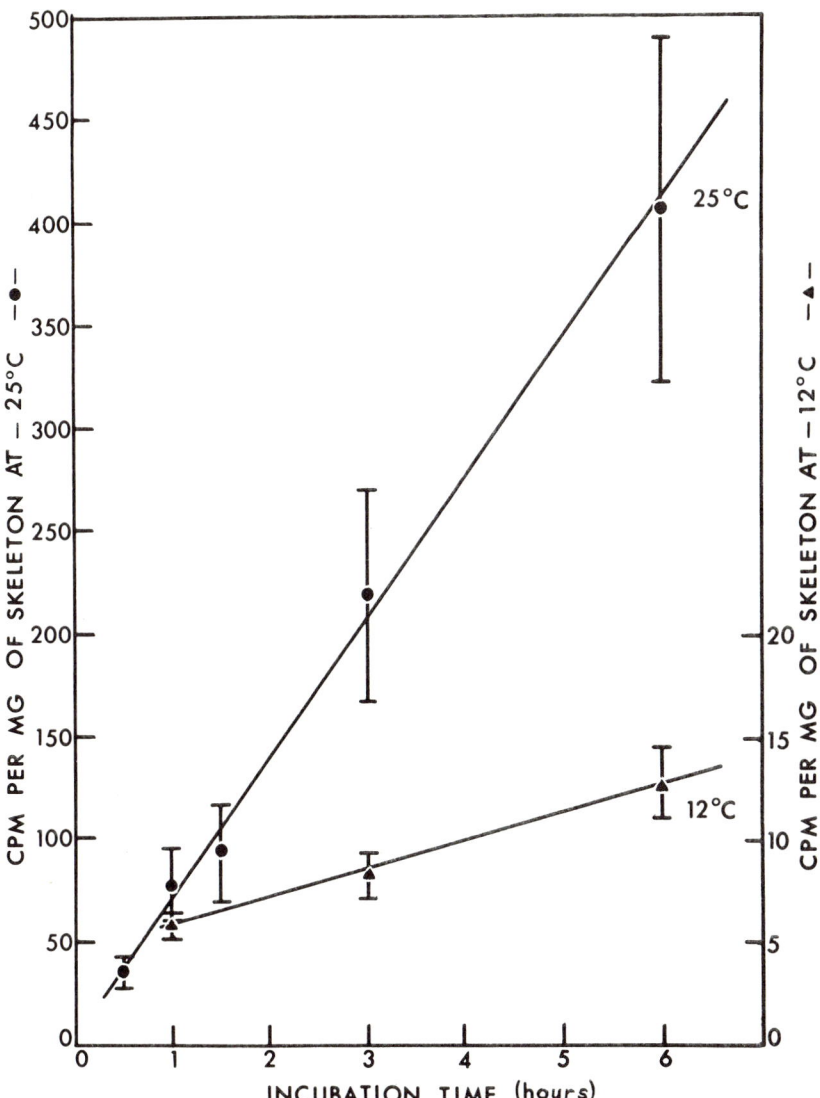

FIG. 5. Amount of ^{45}Ca incorporated by *Pocillopora damicornis* as a function of time at 12° C and 25° C. Note that the scale on the left side (for 25° C) is 10 times as great as the scale on the right side (for 12° C). The vertical bars indicate standard error.

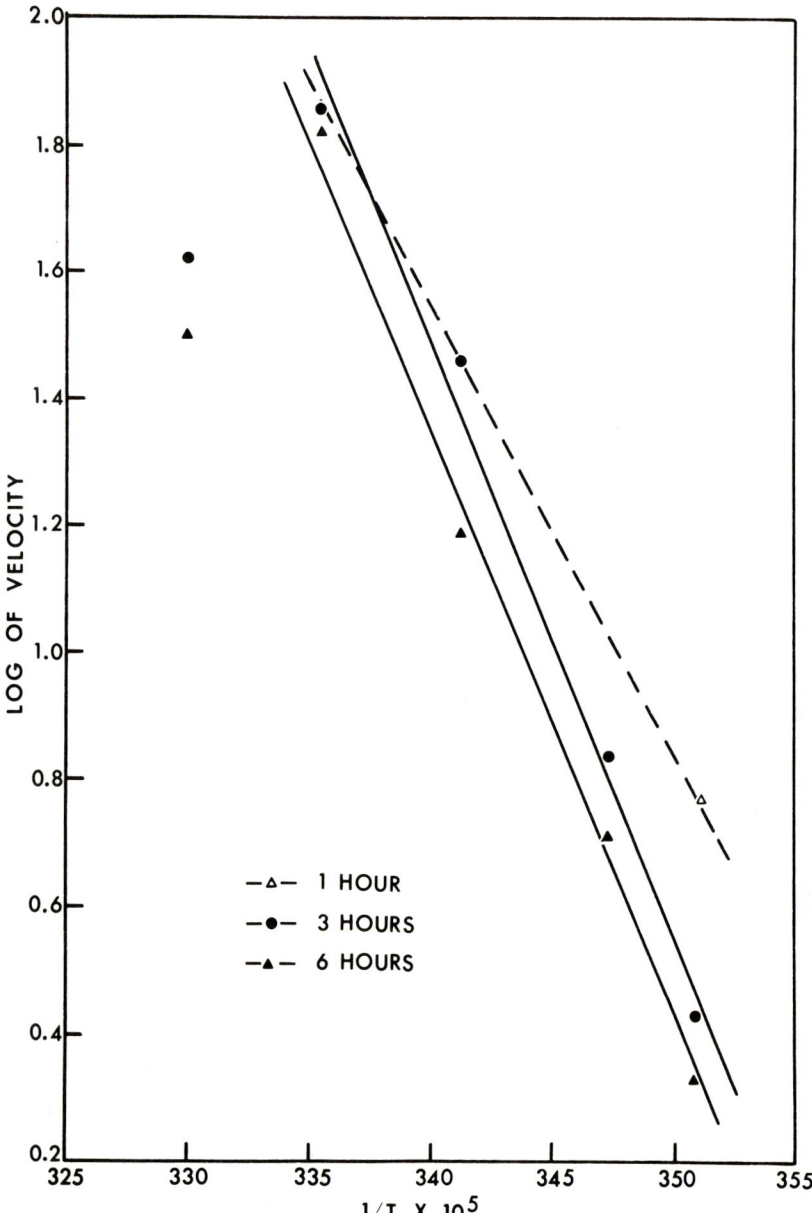

FIG. 6. An Arrhenius plot of the same data as in Fig. 4. The abscissa is the reciprocal of the incubation temperature in degrees absolute. The ordinate is expressed in terms of log of the velocity, because it is assumed that the concentrations of the reacting substances in calcification processes remain constant.

as the concentration of reacting substances probably does not change significantly during the incubation period. Using both the 3- and 6-hour incubation data, two straight lines (excluding the 30° C points) with similar slopes were obtained. The energy of activation (Ea) calculated from the two solid lines (slope equals $-Ea/2.3\ R$) was 43,000 cal/mole. The Q_{10} was 12.7.

Because of the noticeable decrease in calcification rate after 1 hour at 12° C, it was thought that the 1-hour data at 12° C might be a more representative measure of the calcification rate at this temperature. (The lower rate observed at 3 hours and longer may be due to the more advanced stages of "cold death." When this is the case, the assumption that the average velocity is proportional to the initial velocity cannot be extended to those extreme temperatures.) The open-triangle point in Figure 6 represents the 1-hour measurement at 12° C, and the dashed line is based on this measurement rather than on the one made with the 3-hour incubation. This dashed line passes directly through the points obtained for 3-hour incubations at 20° C and 25° C. Unfortunately, no data are available for shorter incubation periods at 15° C. Cold death presents no problem at 20° C and 25° C; hence, these points would be expected to lie on the dashed line. The energy of activation calculated for the dashed line was 33,000 cal/mole and the Q_{10} was 6.7.

DISCUSSION

Technical Suggestions for Studying Calcification Processes in Corals

Despite the advantages of studying an organism under controlled laboratory conditions, problems and sources of error are introduced of which the investigator must be aware in planning and executing the experiment and in interpreting the results. In studies on calcification rates in corals, the general susceptibility of corals to injury from handling and the detrimental effects of certain phases of the experimental procedure on the calcification rate need to be recognized. A brief discussion of some of the problems I encountered in this coral research may be useful to others studying calcification processes in coral.

Mechanical damage to the coral tissues should be avoided by the researcher when he collects and handles the colonies. Furthermore, the colonies should be submerged in seawater as much as possible from the time of collection through all experimental procedures. During the incubation period the possibility of increasing concentration of metabolic waste products (or other materials such as mucus which may inhibit calcification) in the relatively small incubation vessel should be considered in choosing the length of the incubation time. Accumulation of such inhibitory materials may partially account for the consistent decrease in the calcification rates with increasing incubation time (Table 1; Figure 5).

Oxygen depletion and pH changes may be involved. For these reasons the radioactive seawater should be used only once, and it should be prepared just prior to the incubation to eliminate the occurrence of degenerative processes in the seawater. Due to the possible accumulation of products inhibitory to calcification it would be more satisfactory in future experiments to obtain calcification rates based on shorter incubation times. At 25° C satisfactory results were obtained for incubation times as short as 30 minutes.

For the extreme temperatures the shorter incubation periods are even more imperative. The effect of "cold death" or "heat death" on the overall general coral physiology needs to be separated from the effect of changing rates of metabolic reactions, the latter being of primary interest as a cause of changing calcification rates. The bottom row of figures in Table 1 shows that the corals used in the 12° C and 15° C experiments succumbed much more rapidly than those incubated at more moderate temperatures. The calcification rates obtained with the longer incubation times (at the extreme temperatures) thus may reflect not only the slowing down of metabolic reactions vital to calcification, but also degeneration of the general health of the coral by processes associated with either cold or heat death. The processes associated with cold death may explain the rapid decrease in calcification rate taking place after the first hour of incubation at 12° C. Shorter incubation periods would minimize these peripheral or secondary effects and permit more accurate measurements to be made of the primary effects.

On a casual examination of Table 1, a correlation appears to exist between the calcification rates at different temperatures, and the period (in days) during which the coral heads were maintained on the water table before the experiment (Table 1, next to last row). It appears that the greater the number of days the corals were maintained, the lower the calcification rate. Actually there is evidence to indicate that corals kept on the water table did degenerate slowly over a period of days. Although further experiments (see addendum) have shown that no serious error was evidently introduced in this case, it would be well to eliminate this variable in the future by running the experiments on the day the corals are collected.

Parameters for Expressing Calcification Rates

To express ^{45}Ca incorporation or calcification rates in a meaningful way, either a determination of the quantity of coral tissue (responsible for $CaCO_3$ deposition) or some other parameter correlated with tissue quantity is needed. An indirect but rapid method is made possible by the correlation that exists between organic nitrogen of the tissue and skeletal weight or number of calices (Figures 1, 2). Either skeletal weight or number of calices can themselves be directly used as the necessary parameters, or these correlations can give a good approximation of the amount of metabolically active tissue present. The correlation coefficients are greater than .97 for all three graphs (Figures 1–3).

These correlations are linear only in a certain size range; i.e., note that were the lines in Figures 1–3 extrapolated they would not pass through the origin. Because of differing skeletal shapes (and different surface-to-volume ratios), these correlations would have to be made for each coral species separately. Furthermore, because of changing surface-to-volume ratios, other parts of the colony would undoubtedly give different correlations than the branch tips and would need to be determined separately. For those experiments where the conditions of size, shape, and position can be met these correlations provide a rapid method for estimating the organic nitrogen of the coral tissue and suitable parameters for expressing calcification rate.

That such a direct relationship exists between tissue protein and the number of calices was shown by the experiments with hydra (Loomis 1953, 1954; Lenhoff and Loomis 1957; Muscatine and Lenhoff 1965) and with *Cordylophora* (Fulton 1960) which demonstrated that a quantitative indication of the amount of active tissue can be obtained by counting the number of hydranths.

Effect of Temperature on Calcification

Edmondson (1928) showed that hermatypic corals from Hawaii vary in their ability to survive extreme temperatures. Species of *Pocillopora* were particularly sensitive to sudden heating, but had remarkable endurance for rapidly falling temperatures; however, most Hawaiian species of coral, according to his work, were able to survive at least 23 hours at 15° C. At the other extreme, only two out of 13 species of Hawaiian corals survived 32° C for 24 hours. The first corals to succumb, the species of *Pocillopora*, were all dead within 5 hours.

I found that, in those experiments in which the corals were kept at 12° C and 15° C for 6 hours, the animals appeared to be dead the following day; whereas the corals incubated at more moderate temperatures remained healthy in appearance for many days (Table 1). Furthermore, the drop in calcification rate after the first hour at 12° C (Figure 5) probably indicates deterioration of physiological processes due to initial stages of cold death. The decrease in calcification rate above 25° C (at 30° C) also implies some type of deterioration of the general coral physiology due to the high temperature. For convenience then, the determination of calcification rate by temperature can be separated into two effects: (1) the direct effect of temperature on metabolic rate, and (2) the indirect effect of extreme temperature stress on the general health of the coral. In these experiments the latter process becomes especially obvious at 12° C and 30° C, and at the high temperature actually reverses the results that would otherwise be expected.

Because of the indirect effects of stress at these extreme temperatures, the 15° C to 25° C temperature range seems the most useful for studying calcification processes in *Pocillopora damicornis*. In this temperature range the calcification rate shows an exponential increase (Table 1; Figure 4) and an unexpectedly high energy of activation. Activation energies of either 43,000 cal/mole or 33,000 cal/mole are particularly high values; biological reactions are normally around 12,000 to 17,000 cal/mole. A Q_{10} of 12.7 or 6.7 also reflects the particularly large effect of temperature on calcification rate. A Q_{10} of 2 to 4 is more typical of biological reactions.

Since these experiments imply certain upper and lower temperature limits of calcification, together with a striking temperature effect on calcification rate between 15° C and 25° C, it is of interest to consider some observations on the temperature of the water in which coral reefs occur. Vaughan (1919) and Vaughan and Wells (1943) have stated that the minimum yearly ocean temperature to which coral reefs can be exposed and yet survive is between 18° C and 19° C. There may be scattered coral patches in cooler areas but no luxurious reef formation is found there. These temperatues of 18° C–19° C is the range at which a change of a single degree causes a considerable increase in calcification rate (Table 1, Figures 4, 6). Evidently at temperatures below this range, the low rate of calcification cannot build or maintain a reef effectively against normal destruction and dissolution of the $CaCO_3$ that is being laid down. The combined effects then of extreme temperature stress on general coral physiology and the activation energy of calcification reactions are apparently responsible to a considerable extent for luxurious coral reef formations in tropical waters and their sparsity or absence in temperate and frigid waters.

Although an attempt was made to show a physiological basis for the field observations by using the results of these laboratory experiments, the limited applicability of the results should be recognized. Incorporation of ^{45}Ca was measured only in the branch tips of a single coral species from one geographical area. Undoubtedly corals to some extent acclimatize to the temperature environment in which they exist; hence, the temperature limits of survival for a coral species and its calcification rate at a particular temperature would probably vary from one area to another. Different coral species and different parts of the coral colony would also be expected to give different rates. Tips have been shown to be the fastest calcifying area of the coral colony (Goreau 1959). However, if the processes of calcification are similar in all hermatypic corals, then similar activation energies—regardless of the species or its geographical location—might be expected. Further experiments are needed to establish whether this is so or not.

ADDENDUM

More recent work to be reported elsewhere has given results similar to those reported in this chapter. The peak temperature was at 27° C with a rapid decline in rate on both sides. The energy of activation was 41,000 cal/mole—quite similar to that shown in Figure 6. Better experimental techniques eliminated some of the possible sources of error found in the experiments reported in this chapter; however, the effects of the cold death and length of time the corals were maintained on the water table appear not to have introduced the magnitude of error anticipated.

LITERATURE CITED

Edmondson, Charles Howard. 1928. The ecology of an Hawaiian coral reef. *Bernice P. Bishop Museum, bulletin 45.*

Fulton, Chandler. 1960. Culture of a colonial hydroid under controlled conditions. *Science* **132**: 473–474.

Goreau, Thomas F. 1959. The physiology of skeleton formation in corals. I. A method for measuring the rate of calcium deposition by corals under different conditions. *Biological Bulletin, Woods Hole* **116**: 59–75.

Loomis, W. F. 1953. The cultivation of hydra under controlled conditions. *Science* **117**: 565–566.

———. 1954. Environmental factors controlling growth in hydra. *J. Experimental Zoology* **126**: 223–234.

Lenhoff, H. M., and W. F. Loomis. 1957. Environmental factors controlling respiration in hydra. *J. Experimental Zoology* **134**: 171–181.

Muscatine, Leonard, and Howard M. Lenhoff. 1965. Symbiosis of hydra and algae. I. Effects of some environmental cations on growth of symbiotic and aposymbiotic hydra. *Biological Bulletin, Woods Hole* **180**: 415–424.

Vaughan, Thomas Wayland. 1919. Corals and the formation of coral reefs. *Annual Report, Smithsonian Institution* **1917**: 187–276.

Vaughan, Thomas Wayland, and John West Wells. 1943. Revision of the suborders, families, and genera of the Scleractinia. *Special Papers, Geological Society of America* 44.

STEPHEN D. YOUNG
University of California at Los Angeles

Organic Matrices Associated with CaCO$_3$ Skeletons of Several Species of Hermatypic Corals

There have been few studies on the composition of scleractinian coral skeletal matrices. Wainwright (1963), using histochemical and X-ray diffraction evidence, described the skeletal matrix of the reef coral *Pocillopora damicornis* as chitinous, containing little or no protein.

In this paper I present the results of analyses of organic matrices from some Hawaiian corals and compare the data with similar data for the matrices of mollusc shells (Degens, Spencer, and Parker 1967) and brachiopod shells (Jope 1967). The possible influence of the organic matrix on coral calcification is discussed.

MATERIALS AND METHODS

The matrices of the Hawaiian corals *Pocillopora damicornis, Pavona varians, Montipora* sp. (perhaps *Montipora verrilli*), *Cyphastrea ocellina* and *Porites* sp. (perhaps *Porites compressa*) were analyzed.

Isolation of Matrices

Corals were collected and placed in a bucket of tap water overnight. Mucus and superficial tissue were removed with a high pressure stream of freshwater. Skeletons showing green coloration from boring algae were discarded to avoid contamination of the matrix fraction by these algae. Skeletons kept for analysis were placed in 4.8 N KOH at room temperature for at least 24 hours, boiled in fresh 4.8 N KOH for 3 to 5 minutes, then rinsed in running tap water. Skeletons were completely decalcified with 10 percent HCl (approximately 3 N).

The matrix from *Pocillopora damicornis*, which appeared as a white, opaque gel, was collected by passing the solution after decalcification through a Millipore filter (0.45-micron pores). The residue on the filter was washed with distilled water, and the filter and residue were hydrolyzed. Insoluble material remaining after decalcification of *Cyphastrea* and *Pavona* was collected by centrifuging numerous samples of the decalcifying medium at approximately 2,000 rpm for 1 minute in a centrifuge (International Clinical 1530C). *Montipora* and *Porites* skeletons yielded a fine suspension that would clog the Millipore filters before enough material was collected for hydrolysis. To remedy this, the decalcifying solution was mixed with ether and shaken. Droplets formed at the ether-aqueous interface upon which the material precipitated. This material could then be compacted by centrifugation, isolated, and hydrolyzed.

Hydrolysis of Matrices

Each sample of unsoluble residue was transferred to a hydrolysis tube made from 4-mm I.D. glass tubing. About 1 ml of 6 N HCl was added to each tube, the tubes were sealed, and the samples were hydrolyzed for 24 hours at 100° C. The tubes were then centrifuged for 1 minute at about 2,000 rpm to sediment any unhydrolyzed material.

Quantitative Amino Acid Analysis

For analysis in a Beckman amino acid analyzer, matrix hydrolysates were dried under a stream of clean, dry air in a partial vacuum. The dry samples were stored in a desiccator over concentrated sulfuric acid until analyzed. Before analysis, samples were reconstituted in 5 ml of citrate buffer, pH 2.2.

Paper Chromatography

For paper chromatography, matrix hydrolysates were transferred to a porcelain evaporating dish and an equal volume of distilled water added. This solution was then dried over steam. The dry residue was resuspended in water and evaporated to dryness, then suspended in about 0.2 ml of distilled water. The sample was then frozen for later analysis or was chromatogrammed immediately in two dimensions on Whatman no. 4 or no. 1 paper, using phenol:water (72:28) (w/v) as the first dimension and n-butanol:propionic acid:water (1,246:620:884) (v/v/v) as the second dimension (Bassham and Calvin 1957). Amino acids and amino sugars were detected on paper by spraying the paper with 1 percent ninhydrin in acetone and heating the paper at 100° C for 5 minutes. In some analyses the amino sugars were detected by spraying with $AgNO_3$ in acetone and NaOH in 19 percent ethanol after the ninhydrin had faded (Neuberger and Marshall 1966).

RESULTS

Table 1 combines the amino-compound analyses of commercial chitin and the matrices of *Pocillopora damicornis, Montipora* sp., and *Porites* sp. The detection of amino acids and glucosamine after hydrolysis of precipitates has been interpreted as indicating the presence of protein

TABLE 1 Coral matrix hydrolysates

	Asp	Thr	Ser	Pro	Glu	Gly	Ala	Val	Met	Ileu	Leu	Glucos-amine	Tyr	Phe	Orn[†]	Lys	Arg
Commercial Chitin	6	4	5	P*	7	9	10	5	—	1	3	944	1	2	2	—	1
Suborder Astrocoeniina																	
Pocillopora damicornis	22	5	12	—	15	14	7	9	—	—	—	897	6	14	—	—	—
Montipora (verrilli?)	162	38	22	58	111	96	108	94	9	62	106	—	9	56	5	46	18
Suborder Fungiina																	
Porites (compressa?)	220	41	53	49	110	99	74	85	8	59	119	14	19	60	2	60	27

NOTE: Values in residues per 1,000 amino compounds.
*Present in trace amounts.
†Ornithine, created during basic hydrolysis in the reaction: Arg → Orn.
(OH⁻)

and chitin respectively. None of the samples studied contained hydroxyproline or hydroxylysine, indicative of the presence of collagen.

The presence of ornithine is interpreted as resulting from the alkaline hydrolysis of arginine. Thus, the ratio of ornithine to arginine indicates the extent of alkaline hydrolysis that might have occurred while the coral was being cleaned with KOH. That the ornithine:arginine ratio is low for *Montipora* and *Porites,* compared to commercial chitin, may indicate that alkaline cleaning did not cause significant hydrolyses of matrices.

It should be noted that the composition of *Montipora* and *Porites* matrices are similar, but *Porites* is easily distinguished because it contains glucosamine and *Montipora* does not. The chitinous matrix of *Pocillopora* was not purified between decalcification and hydrolysis, yet, as Table 1 shows, it is almost as pure as the commercially prepared, deproteinized arthropod chitin.

Results similar to those described in Table 1 were obtained with paper chromatography for *Pocillopora, Montipora,* and *Porites.* The matrices of three other species of coral were examined by paper chromatography alone. These were *Pavona varians, Cyphastrea ocellina,* and *Fungia scutaria. Pavona* showed only one ninhydrin-positive spot which was not analyzed further. The matrix from *Cyphastrea* also showed a single ninhydrin-positive spot. This has been tentatively identified as threonine. The *Fungia* matrix, like that of *Pocillopora,* contained mainly glucosamine.

DISCUSSION

The analyses of *Pocillopora* matrix (Table 1) confirm Wainwright's finding (1963) that chitin is the major constituent. The diversity and quantity of amino acids detected suggests that one or more protein components are associated with the matrix. No attempt was made to determine whether any amino acids are covalently bond to the chitin. Inasmuch as the amino acids were not removed when the isolated matrix was washed, I have assumed that they are bound to the chitin in some way or that they come from one of the filamentous fractions of *Pocillopora* skeletal matrix described by Wainwright (1963). The association of chitin and protein in the insect cuticle is described by Hackman (1960).

Of the corals examined, only *Pocillopora* and *Fungia* had chitinous matrices. The *Porites* matrix yielded some glucosamine, but it is not clear whether this represents a small amount of associated chitin or part of some other polysaccharide. Only amino sugars are detected by the method used here; other sugars, if present, would have been undetected.

The detection of only one ninhydrin-positive spot on chromatograms of hydrolyzed matrices of *Pavona* and *Cyphastrea* suggests that the matrices of these corals may differ from those described in Table 1, and that they may be rich in a single amino acid. Further analyses of larger amounts of these matrices would prove interesting.

The lack of similarity in matrix composition in these corals is surprising. Work on skeletal matrices in molluscs and brachiopods suggests that matrix composition differs only slightly between closely related forms (Degens et al. 1967; Jope 1967). Considering the data in Table 1 and the results for *Fungia* (also a fungiid, but containing a large proportion of glucosamine), I can discover no correlation between the chemical composition of the various coral matrix types and the taxonomic classification of the coral.

This lack of similarity in the chemical composition of coral matrices is also important when considering the mechanisms of calcification. Wilbur (1960) asserted that the protein component in molluscan matrices controls some aspects of calcification. Wainwright (1963) hypothesized that the presence of amino groups may control deposition of aragonite. Since all coral matrices studied have amino acids and some have amino sugars, the variation seen does not preclude the possibility that the amino groups present, or even specific amino acids, are important to mineralization. The wide variation of matrix types in corals may also mean that the organic matrix is unimportant in calcification, or that matrix may suppress calcification by functioning as a crystal poison. Whatever the explanation, coral matrix constituents vary widely from species to species.

LITERATURE CITED

Bassham, J. A., and M. Calvin. 1957. *The path of carbon in photosynthesis.* Englewood Cliffs, N.J.: Prentice-Hall, Inc.

Degens, E. T., D. W. Spencer, and R. H. Parker. 1967. Paleobiochemistry of molluscan shell proteins. *Comparative Biochemistry and Physiology* **20**: 533–579.

Hackman, R. H. 1960. Studies on chitin. IV. The occurrence of complexes in which chitin and protein are covalently linked. *Australian J. Biological Sciences* **13**: 568–577.

Jope, M. 1967. The protein of brachipod shell. I. Amino acid composition and implied protein taxonomy. *Comparative Biochemistry and Physiology* **20**: 593–600.

Neuberger, A., and R. D. Marshall. 1966. Methods of qualitative and quantitative analysis of the component sugars. In *Glycoproteins: Their composition, structure and function,* A. Gottschalk, ed., pp. 190–234. New York: Elsevier Publishing Co.

Wainwright, S. A. 1963. Skeletal organization in the coral, *Pocillopora damicornis. Quarterly J. Microscopical Science* **104**: 169–183.

Wilbur, K. M. 1960. Shell structure and mineralization in molluscs. In *Calcification in biological systems,* R.F. Soggnaes, ed., pp. 15–40. Washington, D.C.: American Association for the Advancement of Science.

HOWARD M. LENHOFF APPENDIX
University of California at Irvine
BERTON ROFFMAN
College of the Holy Cross, Worcester, Massachusetts

Two Methods for Fractionating Small Amounts of Radioactive Tissue

Many methods have been devised for fractionating tissue into its various major classes of chemical components. We describe two such methods which were developed for hydra, and which are especially applicable to small amounts of radioactive animal tissue.

MATERIALS

Aside from the usual laboratory supplies, for these fractionations the following materials were necessary: trichloroacetic acid (TCA), 5 percent and 10 percent solutions; micropipettes; Millipore filters, 0.45-micron pore size; microanalysis filter holder (Millipore Filter Corp., XX 10 100 00); constant-temperature water baths; equipment for counting radioactivity, either gas-flow or liquid scintillation.

METHOD I

This method, which was the chief one employed by the students in this program, makes use of Millipore filters primarily to eliminate a major source of error that usually occurs when working with small amounts of tissue. When one separates soluble from insoluble components by centrifuging small volumes of a mixture, some particles of the precipitate often dislodge from the tube and contaminate the soluble fraction. Although the actual amount of precipitate that dislodges may be minute, its radioactivity may significantly alter the results. Hence, before being counted, all solutions were first passed through a Millipore filter. The filters were also used to eliminate the last centrifugation step in the fractionation procedure; instead, in the last step, the soluble material was separated from the insoluble material by filtration. This technique is exploited to the fullest in Method II in which filtrations are substituted for centrifugations throughout. Millipore filters work best for such separations when small amounts of tissue (e.g., 10 hydra) are being studied. Larger amounts frequently clog the pores of the filter.

Fractionation Procedures

In this method, 10 labeled hydra are usually sufficient for a fractionation. Before the fractionation, the animals are washed in clean culture solution and are homogenized with either a small glass tissue grinder or by a sonic oscillator.

Fraction 1. From a 1-ml suspension of homogenized tissue, remove 100 µl for counting.

Fraction 2. To the remaining 0.9 ml of homogenate, add 0.9 ml of 10-percent TCA, let the mixture stand at room temperature for 15 minutes and centrifuge at about 10,000 X g for 15 minutes. Pass all the supernatant through a Millipore filter and collect the TCA-soluble material in a clean test tube. Extract the TCA from this solution with three 2-ml portions of ether.

Fraction 3. Wash the "TCA particles" remaining on the filter three times with 10-ml portions of 5 percent TCA.

Fraction 4. To the TCA-insoluble precipitate remaining in the centrifuge tube, add 4 ml of 80 percent ethanol and one drop of 0.1 N HCl. Heat the tube in a 45° C water bath for 30–40 minutes, and stir the contents occasionally. Centrifuge the tube and its contents at 10,000 X g for 15 minutes. Pour the supernatant through a Millipore filter into a clean test tube to get the TCA-insoluble, alcohol-soluble material. (Depending upon the label used and the amount of lipid material labeled, the remaining precipitate can be extracted again with a 1:1 ratio of 80 percent ethanol:ether.)

Fraction 5. Wash the "alcohol particles" remaining on the filter three times with 10-ml portions of 80 percent ethanol.

Fraction 6. To the alcohol-insoluble precipitate remaining in the centrifuge tube, add 4 ml of 5 percent TCA and heat the resultant suspension in a boiling water bath for 30 minutes. Cool the tube and pass 1 ml of the suspension through a Millipore filter and collect the TCA- and alcohol-insoluble, hot TCA-soluble material in a clean tube. Extract the TCA as in fraction 2.

Fraction 7. With three portions of 10 ml 5 percent TCA, wash the filter pad containing the TCA-, alcohol-, and hot TCA-insoluble material.

Assay

The filter pads with radioactive material from fractions 3, 5, and 7 can be counted directly. The volume of the soluble material counted depends upon the type of label and its distribution among the fractions. We usually sample 200-μl portions from fraction 2, 1 ml from fraction 4, and 0.4 ml from fraction 6.

Calculations

Calculations of the total radioactivity in a particular chemical fraction are complicated. Such calculations not only take into account the obvious dilution and size of sample factors (discussed more fully with examples under Method II), but they also have to account for the radioactivity in the dislodged particles that were salvaged in fractions 3 and 5. Fraction 5 contains both material which is soluble and material which is insoluble in hot TCA (i.e., from fractions 6 and 7), whereas fraction 3 not only contains these, but in addition, some alcohol-soluble material (as in fraction 4). Thus, once the percent distribution of radioactivity for fractions 2, 4, 6, and 7 has been determined, the relative amounts of 4, 6, and 7 in fraction 3 are calculated, and the amounts of 6 and 7 in fraction 5 are calculated. These values are then added to those of the proper major fraction.

An example of the above is illustrated as follows. Let x = the counts in fraction 5, y = those in 6, and z = those in 7. Thus the counts of fraction 6 represented in the counts of the particles of fraction 5 would be $[y/(y + z)] \cdot x$. This value is then added to the counts of fraction 6 that were originally counted. When the same kind of calculation has been done for the particles of fraction 3, and the dilution factors have been corrected, then the final percent distribution can be totaled.

Final distribution of label will be expressed as four major fractions: (*a*) TCA-soluble, containing mostly small molecules and sugars; (*b*) TCA-insoluble, alcohol-soluble, containing lipids, lipid-soluble materials and small proteins; (*c*) TCA- and alcohol-insoluble, hot TCA-soluble, containing nucleic acid components; and (*d*) TCA- , alcohol- , and hot TCA-insoluble, containing mostly protein.

METHOD II

This method differs from previous methods in two regards. (1) Rather than separating soluble from insoluble material by centrifugation steps, separation is achieved entirely through use of Millipore filters. In essence, this method consists of treating different portions of the suspended material with one of a series of reagents. By simply filtering each of these mixtures, counting the filter pad and some of the filtrate, and making a few calculations, the distribution of radioactivity in all the fractions can be determined. (2) This method allows the separation of polysaccharides from monosaccharides in the TCA-soluble fraction.

Fractionation Procedures

This method, although more rapid than Method I, requires a larger initial sample of tissue. Using hydra, and depending upon the amount of radioactivity, usually 50–200 animals homogenized in a 3-ml volume will do.

Fraction 1. This is a 0.4-ml sample of the total initial suspension.

Fractions 2 and 3. To 1 ml of suspension, add 1 ml of 10 percent TCA and allow the mixture to stand at room temperature for about 15 minutes. Filter 1.5 ml of the now 5 percent TCA mixture, and collect the filtrate in a clean test tube. The TCA-soluble material is marked fraction 2, and the TCA-insoluble material on the filter is fraction 3. Wash fraction 3 on the filter pad with three 10-ml portions of 5 percent TCA. (See next steps for obtaining fractions 8 and 9.)

Fractions 4 and 5. Repeat the above step, only this time heat the mixture for 30 minutes in a boiling water bath after adding the 10 percent TCA. Separate through the filter into the hot TCA-soluble (4) and hot TCA-insoluble (5) fractions. Wash fraction 5 as above.

Fractions 6 and 7. To 0.6 ml of the suspension, add 3.2 ml of 95 percent ethanol, and incubate at 45° C for 45 minutes. Remove 3 ml of the stirred suspension and pass it through the Millipore filter. An alcohol-soluble (6) and an alcohol-insoluble (7) fraction result. Wash fraction 7 with three portions of 80 percent ethanol.

Fractions 8 and 9. To 0.6 ml of the TCA-soluble material from fraction 2, add 3.2 ml of 95 percent ethanol and heat at 45° C for 45 minutes. Remove 3 ml of the stirred suspension and filter into the TCA-soluble alcohol-soluble fraction (8) and the TCA-soluble alcohol-insoluble fraction (9). Wash fraction 9 with three portions of 80 percent ethanol.

Assay

The filter pads with radioactive material from fractions 3, 5, 7, and 9 can be counted directly. The volume of the soluble material counted depends upon the distribution and type of label. For example, in the fractionation listed in Table 2 A, B, we removed 0.4 ml from fractions 1, 2, and 4, and 1 ml from fractions 6 and 8.

Calculations

Calculations for the total radioactivity in a particular sample have to take into account the dilution of the sample in each procedure and the size of the sample counted. For example, to determine the amount of activity in the original homogenate represented by fraction 4, three adjustments must be made: only one-third of the original homogenate is used, only 1.5 ml of the 2-ml extraction mixture is filtered, and only 0.4 ml of 1.5 ml of the filtrate is assayed for radioactivity. Thus the factor used to correct for these three steps is derived from the reciprocals of the dilution factors: i.e., 3/1 X 2/1.5 X 1.5/0.4 = 15. Dilution factors are calculated for all fractions in this manner.

The percentage of radioactivity in the major biochemical fractions of the original radioactive tissue analyzed can be easily calculated from the radioactivity found in each of the filter fractions. Table 1 lists the major fractions, the possible components of each, and the corresponding filter fraction(s) from which the radioactivity in the major fraction was derived. Data from a sample fractionation are presented in Table 2A, B.

DISCUSSION AND CONCLUSIONS

The methods as presented herein are only two of the many modifications of these general procedures now in use. We have presented them only in their most general forms. Even in this summer program, the students modified the procedures for their own particular needs. For example, because Cook and DiSalvo used only ^{35}S-labeled material, they did not expect label to be in the fractions containing nucleic acids. Hence they eliminated the steps for fractions 5, 6, and 7 of Method I and substituted a filtration for the last centrifugation step in separating the alcohol-soluble from the alcohol-insoluble fractions. Murdock, in his experiments on sea anemones, used ^{14}C-labeled material. Hence, he carried out the complete fractionation. Gosline added an autoclaving step to separate collagens and protocollagen from the other proteins.

Method II was formulated after the course. It was developed when we found that green hydra, given $^{14}CO_2$ in the light, concentrated most of the animal tissues' label in the TCA-soluble fraction. This unusually high TCA-soluble fraction prompted us to dissect this fraction further into its alcohol-soluble and alcohol-insoluble components. No doubt the use of other labels and the running of experiments under different physiological conditions will lead to further modifications of these methods.

TABLE 1 Composition of major classes of chemical components

Major Fraction		Probable Components	Corresponding Fraction of Method II
A	(TCA-soluble, ethanol-soluble)	Small molecules, such as amino acids and monosaccharides	8
B	(TCA-soluble, ethanol-insoluble)	Oligosaccharides, oligonucleotides	9
C	(TCA-insoluble, ethanol-soluble)	Lipids and lipid-soluble compounds; small proteins	6–8
D	(TCA- and ethanol-insoluble, hot TCA-soluble)	Nucleic acid components	4–2
E	(TCA-, ethanol-, and hot TCA-insoluble)	Protein	5–(6–8)

TABLE 2A Example of sample analysis by Method II

Fraction	cpm	Dilution Factors	Corrected cpm in Total Sample
1	1,961	7.5	14,708
2	75	15	1,125
3	3,412	4	13,648
4	289	15	4,335
5	2,741	4	10,964
6	82	19	1,558
7	2,133	16.3	13,438
8	26	38	988
9	20	12.6	252

TABLE 2B Final tabulation of counts in major fractions using data of Table 2A

Major Fraction	Corrected cpm	Percent
A	988	6.4
B	252	1.6
C	570	3.7
D	3,210	20.8
E	10,394	67.4
TOTAL	15,414	

NOTE: Percent recovery = sum of counts observed in all fractions divided by counts in fraction 1 times 100. In this case, 105 percent of the label in the original sample was accounted for in the fractions analyzed.

Index

Acaulis ilonae, 17 tab 1; food requirements of, 22
Acontiate sea anemone, feeding responses of, 77 tab 1
Acrophores, 163, 166 fig 2
Acropora cervicornis, 232
Acroporidae, 49
Actinae, 80
Actinia mesembryanthemum, presence of actinohaematin in, 80
Actiniaria, 45–47, 49, 173 tab 1
Actinohaematin: in *Actinia mesembryanthemum,* 80; in *Bunodes crassicornis,* 80
Aequorea aequorea, 154
Aequorea coerulescens, 18 tab 1
Agariciidae, 49
Aiptasia sp., 45, 46 fig 11, 49, 173 tab 1, 174; *Artemia salina* as food for, 137, 141, 146, 147, 169, 218, 219; collagen in, 148, 153–155; collection and maintenance of, 136, 146, 147, 169, 218, 219; feeding of labeled food to, 141, 144, 145; feeding of ^{35}S amino acids to, 218–224; ^{14}C proline-^{14}C hydroxyproline ratios in, 153 fig 5, 154, 155; fractionation of, 141, 142, 143 fig 2, 144, 147–150, 152, 154, 219, 220; G6PDH and 6PGDH activity in, 173 tab 1; hydroxyproline assay in, 148, 149, 152; mesogleal collagen formation in, 146, 149, 153, 154; proline assay in, 148, 149, 152; radioactivity in, 142 fig 2, 149 fig 1, 150 fig 2, 151 figs 3–4, 152 fig 6; response to ^{14}C-proline of, 146–148, 149 fig 1, 154; uptake of ^{35}S in, 221 fig 1, 222 tabs 1–2
Aiptasia pallida: collagens in, 80, 157; nematocysts of, 159, 162
Aiptasia pulchella, effect of DTE on, 161 tab 1
Alcyonacea, 187, 228 tab 1
Alcyonium digitatum, 129
Algae, 219; coelenterate associations with, 182, 192; endosymbiosis of, 179–187; invertebrate symbiosis of, 218; radioactivity in, 221 fig 1
Amastigophores, 164; macrobasic, 164, 167 fig 14; microbasic, 164, 167 fig 13
Amino acids, 85, 89, 209, 261; in coelenterates, 76 tab 1; feeding response of *Boloceroides* sp. to, 95 tab 1; feeding response of *Cyphastrea ocellina* to, 102, 105, 114, 115; radioactivity in, 78, 79
Amphinema dinema, 17 tab 1
Amphinema rugosum, 17 tab 1
Anacrophores, 163, 166 fig 1
Anemonia sulcata, 181, 185
Anion requirements, for hydras, 10
Anisorhiza, 164; heterotrichous, 164, 166 fig 10; homotrichous, 164, 166 fig 9
Anthomedusae (= Gymnoblastea), 17 tab 1

Anthopleura sp., 185, 186; effect of DTE on, 161 tab 1; G6PDH and 6PGDH activity in, 173 tab 1
Anthopleura elegantissima, 184
Anthopleura nigrescians, G6PDH and 6PGDH activity in, 173 tab 1
Anthozoa, 160 tab 1, 173 tab 1, 228 tab 1
Artemia salina, 22, 23, 60, 61, 96, 119, 126; chromatography of, 95; extract of, 86; as food for *Aiptasia* sp., 137, 141, 146, 147, 218, 222; as food for *Boloceroides* sp., 92, 93, 95, 169; as food for *Chlorohydra viridissima,* 193–196; as food for *Cyphastrea ocellina,* 101, 103, 105 fig 3, 115; as food for *Fungia scutaria,* 240; as food for *Hydra littoralis,* 116, 117; as food for laboratory-raised hydra, 12, 13, 20, 21; as food for *Pennaria tiarella,* 86, 87, 91, 119, 120, 124; as food for *Pocillopora damicornis,* 67, 69, 248; as unacceptable food for *Zoanthus sandwichensis,* 209; fractionation of, 95
Ascaris sp., collagens in, 81
Astomocnides, 163
Aurelia sp., feeding responses of, 77 tab 1
Aurelia aurita, laboratory culture of, 9
Autoradiography, of *Pennaria tiarella,* 120, 123, 124 fig 2

Bacteria: analysis of *Zoanthus sandwichensis,* 211; in aquatic ecosystems, 129; ingestion by *Fungia scutaria* of, 129, 131, 133; ingestion by *Montipora verrucosa* of, 133; ingestion by *Pavona varians* of, 133; ingestion by *Pocillopora damicornis* of 129, 133; ingestion by *Pocillopora meandrina* of, 133; ingestion by *Porites compressa* of, 133; in studies of corals, 129–135
Birhopaloides, 165, 167 fig 22
Boloceroides sp., 159, 174; *Artemia salina* as food for, 92, 93, 95, 169; changes in volume of, 93; collection and maintenance of, 92, 169; feeding responses of, 77 tab 1; G6PDH and 6PGDH activity in, 173 tab 1; "Instant Ocean" as medium for, 92; response to amino acids of, 95 tab 1, 96, 114; response to live food of, 93, 94 figs 1–2; response to valine of, 89, 96, 97 tab 2, 98, 100
Boloceroides lilae, 45, 47, 49; effect of DTE on, 160 tab 1; G6PDH and 6PGDH activity in, 173 tab 1
Bougainvillia sp.: artificial seawater formula for, 53 tab 1; description of early growth in laboratory of, 17; effects of dilution of seawater upon, 58 tab 4, 59 tab 5; effects of salinity on, 55, 56; influence of quantity of food upon, 22; ionic requirements of, 57, 60 fig 2, 63 tab 8, 64; laboratory growth of, 53, 54 fig 1, 55, 56 tab 2, 57 tab 3, 63; nutritional levels for, 61, 62 tab 7; temperature tolerance of, 61 tab 6, 64
Bougainvillia carolinensis, 17 tab 1; effects of temperature on stolon growth of, 32; growth rates of, 62; influence of quantity of food upon, 22; stolon growth rate of, 27
Bougainvillia superciliaris, effect of temperature on growth of, 64
Bunodes crassicornis, presence of actinohaematins in, 80

Calcium carbonate: precipitation in corals of, 227–236; sources of carbonates in, 239, 255, 256
Calicoblast, cells in corals, 227, 228
Campanularia calceolifera, 18 tab 1
Campanularia flexuosa, 18 tab 1, 126; effect of glutathione on, 92, 100; feeding responses of, 77 tab 1; food distribution in, 119; hydranth development in, 24; influence of quantity of food upon, 22; mitosis in, 27; normal reproduction of, 20; stolon growth in, 25, 27, 30
Carybdea sp., 174 tab 2; G6PDH and 6PGDH activity in, 172 tab 1

Index

Cassiopeia sp., 174 tab 2, 181, 182, 184, 186; effect of DTE on, 160 tab 1; G6PDH and 6PGDH activity in, 172 tab 1

"CCS-5," growth medium for *Cordylophora lacustris*, 23, 170

Ceriantharia, 49

Cerianthus sp., 49

Chlorella sp., 179

Chloride ions: in *Cordylophora lacustris*, 10; in *Hydra littoralis*, 10

Chlorococcales, description of, 179

Chlorohydra chlorella, 179–181

Chlorohydra viridissima, 171, 174 tab 2, 175; *Artemia salina* as food for, 193; control of sexual differentiation in, 9; fixation of $^{14}CO_2$ in, 205, 206 tab 2; fractionation of, 203, 205; gonadogenesis in, 11; growth rate of, 62; G6PDH and 6PGDH activity in, 172 tab 1; influence of quantity of food upon, 22; laboratory culture of, 193, 202; magnesium ions in, 10; potassium ions in, 10; response to $^{14}CO_2$ of, 192–200, 202 figs 1–5, 203, 204 tab 1, 205; response to glutathione of, 116, 117; sodium ions in, 10; utilization of metabolic CO_2 by, 192

Chromatography, 95, 131, 210, 213 fig 1, 261

Cladonema radiatum, 17 tab 1

Clava multicornis, 17 tab 1; stolon growth in, 25

Clytia johnstoni, 18 tab 1; influence of quantity of food upon, 22, 24 fig 4; mitosis in, 27, 28; stolonal activity in, 29, 30

Cnidocyte, migration in hydra, 9, 11

Coelenterates: amino acid activity in, 76 tab 1, 98, 100; *Artemia salina* as food for, 169; asexual reproduction of, 13; chemistry of digestion of, 78–80; collection and maintenance of, 157, 158; current research on, 75, 76; disulphide linkages in, 157; early research on, 9, 75; effect of valine on, 98; endosymbiosis in, 179–187; enzyme pathways in, 81; G6PDH and 6PGDH activity in, 169–175; husbandry of, 9–11; intercellular digestive chemistry of, 79, 137; laboratory rearing of, 11, 113; mesogleal collagen of, 146, 154, 155; metabolism and biochemistry of, 79, 80; nematocyst treatment with DTE of, 157–162; protein composition of, 157; skeletal calcification in, 236; skeletal types among, 228 tab 1; symbiosis of algae with, 179, 192; zoochlorellae in, 179, 180; zooxanthellae in, 181–183, 188

Coenothecalia, 228 tab 1

Collagen: in *Aiptasia pallida*, 80, 81, 148, 153–155; in *Ascaris* sp., 81; in coelenterates, 146, 154, 157; disulphide bonds in, 157; in hydras, 80

Corals: bacteria studies in, 129–135; calcification processes in, 255, 256; calcium carbonate precipitation in, 229, 230; calicoblast cells in, 227, 228; crystal formation in skeleton of, 234, 235; micromorphology of skeleton of, 299–232; mineralogy of skeleton of, 234, 235; morphology of tissues of, 227; physiology of calcification in, 233, 234; site of calcium carbonate deposition in, 227; temperature effects on, 246

Cordylophora lacustris, 77 tab 1, 126, 170; *Artemia salina* as food for, 20, 21; chloride ions in, 10; feeding response to pipecolic acid of, 89, 107; feeding response to proline of, 84, 89, 91, 92, 97, 100, 114; feeding responses of, 76, 77 tab 1; growth in "CCS-5," 23; growth rate of, 62, 63; G6PDH and 6PGDH activity in, 172 tab 1; influence of quantity of food upon, 22; laboratory culture of, 9; magnesium ions in, 10; nitrate ions in, 10; sensitivity to changes in culture solution of, 12, 17; stolon growth rates of, 25–27, 29

Corymorpha palma: influence of quantity of food upon, 22; stolonal growth in, 30
Coryne tubulosa, 17 tab 1
Craspedacusta sowerbii: formation of frustules in, 30, 31; sensitivity to changes in culture solution of, 12, 18 tab 1
Ctenophora, 173 tab 1
Culture solution, 11, 12; development of, 13; importance of composition of, 11
Cydippida sp., G6PDH and 6PGDH activity in, 173 tab 1
Cyphastrea ocellina, 49, 159; amino acid analysis of, 261; analysis of organic matrix of, 260; *Artemia salina* as food for, 101, 109; chromatography of, 261; collection and maintenance of, 101, 102, 260; effect of DTE on, 161 tab 1; feeding responses of, 77 tab 1; hydrolysis of matrix of, 261, 263; isolation of matrix of, 260, 261, 263, 264; mouth-opening behavior of, 106 tabs 1–2, 114; plankton as food for, 101; response to glutathione of, 107, 109, 113 tab 6, 114, 115; response to hydroxyproline of, 107 fig 4, 108; response to live prey of, 102, 103 fig 1, 105 fig 3, 115; response to pipecolic acid of, 100, 107, 114; response to proline of, 90, 97, 107–110, 114
Cytochrome, 79, 80

Dendrophyllia manni (= *Tubastrea aurea*), 45, 49; G6PDH and 6PGDH activity in, 173 tab 1
Dendrophyllidae, 45, 49
Desmonemes, 163, 166 fig 3
Diadumene sp., 159
Dictyosphaeria intermedia, 218
Dipurena halterata, 17 tab 1
Disulphide bonds: in coelenterates, 157; in collagens, 157

Dithioerythritol (DTE), effects on coelenterates of, 157–162
DTE. *See* Dithioerythritol

Eleutheria dichotoma, 17 tab 1
Endosymbiosis, of algae, 179–187
Enzyme pathways: in coelenterates, 81; in *Hydra littoralis,* 81
Epiactis prolifera, feeding responses to chemical activators, 98
Escherichia coli, 138
Eurythele: heterotrichous microbasic, 164, 167 fig 16; holotrichous macrobasic, 165, 167 fig 20; homotrichous microbasic, 164, 167 fig 15; merotrichous macrobasic, 165, 167 fig 19; semiphoric microbasic, 164, 167 fig 17; teleotrichous macrobasic, 165, 167 fig 18

Faviidae, 49
Feeding: chemical control of, 11; of hydra, 10; response to cyanide of, 9
^{45}Calcium, uptake by *Pocillopora damicornis* of, 246, 248–259
^{14}C-glycine, uptake by *Zoanthus sandwichensis* of, 209, 213–215, 216 tab 4
^{14}C-hydroxyproline, activity in *Aiptasia* sp., 148, 154, 155
^{14}CO$_2$: effect of food on uptake of, 194 fig 1, 195 fig 2, 196 fig 3, 197 fig 4; uptake by *Chlorohydra viridissima* of, 192–207 tabs 1–2; variation of fixation in hydra of, 196, 198 fig 5, 206 tab 2
^{14}C-proline, activity in *Aiptasia* sp., 146–148, 150, 154, 155
Fractionation: of *Aiptasia* sp., 141, 142, 143 fig 2, 144, 147–150, 152, 154; of *Artemia salina,* 95; assay of, 267; calculation of radioactivity in, 267, 269; of *Chlorohydra viridissima,* 203, 205; of *Fungia scutaria,* 131,

135, 240, 241, 244; of *Hydra littoralis*, 138, 139, 144; materials for, 265; methods of, 265–271, 270 tab 1, 271 tabs 2A–2B; of *Pocillopora damicornis*, 131, 135; of *Zoanthus sandwichensis*, 210
Fungia fragilis, 38, 40 fig 4, 49
Fungia (hexagonalis ?), 41 fig 5
Fungia patella, 38, 40 fig 3, 49
Fungia scutaria, 38, 39 figs 1–2, 49, 115, 173 tab 1; *Artemia salina* as food for, 110, 240; chromatography of, 131; collection and maintenance of, 129, 130, 239, 240; distribution of radioactivity in, 241 tab 1, 242 tab 2; effect of DTE on, 161 tab 1; feeding mechanisms of, 135; feeding of ^{14}C-labeled mouse tissue to, 239, 240, 244; feeding of labeled bacteria to, 130, 131, 132 tab 1, 133, 134 tab 2, 135; feeding response to glutathione of, 110, 113 tab 6, 116; feeding response to proline of, 104 fig 2, 110, 116; ^{14}C-amino acid labeling of, 243, 244; $^{14}CO_2$ incorporation of, 239, 243, 244; fractionation of, 131, 135, 240, 241, 244; G6PDH and 6PGDH activity in, 173 tab 1; hydrolysis of, 263; mouth-opening behavior of, 110, 111 tab 4; mucous secretion in, 110, 132
Fungiidae, 38–40, 49

Galaxea sp., 230
Glucose-6-phosphate dehydrogenase (G6PDH), 169–173 tab 1, 174, 175; fluorometric measurements of, 170
Glutathione, 139; as feeding activator of hydras, 9, 76, 84, 98, 100; feeding response of *Campanularia flexuosa* to, 92, 100; feeding response of *Chlorohydra viridissima* to, 116, 117; feeding response of *Cyphastrea ocellina* to, 100, 105, 107, 108 fig 5, 109 tab 3, 114–116; feeding response of *Hydra littoralis* to, 92, 97, 116; feeding response of *Hydra pirardi* to, 117; feeding response of *Pennaria tiarella* to, 88, 89 tab 2
Gonadogenesis, in *Chlorohydra viridissima*, 11
Gorgonacea, 187, 228 tab 1
Growth media, for colonial hydroids, 22, 23
G6PDH. *See* Glucose-6-phosphate dehydrogenase

Hapalocarcinus marsupialis, 45
Haplonemes, 163
Heterotrichous, 164
Histohaematins. *See* Cytochrome
Hydra littoralis, 12, 119, 120, 169, 171, 199; *Artemia salina* as food for, 116, 117; chloride ions in, 10; collection and maintenance of, 137, 138; control of sexual differentiation of, 9; effect of glutathione on, 84, 92, 97, 100, 116; enzyme pathways in, 81; feeding of labeled food to, 138, 139, 143, 145; feeding study of, 138, 139, 144; growth rate of, 62; G6PDH and 6PGDH activity in, 171; influence of quantity of food upon, 22; mouth-opening behavior of, 88; nematocyst capsules of, 157; normal reproduction of, 20; potassium ions in, 10; production of $^{14}CO_2$ in, 192; radioactive food analysis in, 138, 139, 140 fig 1, 143, 145; sodium ions in, 10
Hydra pirardi, feeding response to glutathione, 117
Hydra pseudo-oligactis, 174 tab 2; G6PDH and 6PGDH activity in, 172 tab 1
Hydractinia echinata, 17 tab 1; abnormal growth of, 20; dietary supplements for, 20; normal reproduction of, 20; stolon growth in, 27
Hydranth development: factors influencing, 24; in *Campanularia*

flexuosa, 24; in colonial hydroids, 19 tab 2, 24; in *Pennaria tiarella,* 86
Hydras: abnormalities in, 11; anion requirements of, 10; *Artemia salina* as food for, 12; budding changes in, 12; cnidocyte migration in, 11; collagens in, 80; culture techniques for, 180; current research of, 75, 76; defense mechanisms of, 10; effect of glutathione on, 9; environmental factors of, 13; feeding behavior of, 10, 11, 76, 77 tab 1; growth and maintenance of, 10; intracellular digestive biochemistry of, 10; metabolic rate of, 10; methods of culture of, 9; mutants of, 11; surface attachment of, 12; symbiosis of algae with, 10, 180, 181; use of radioactive amino acids on, 79
Hydroida, 48–49, 172 tab 1
Hydroids, colonial: G6PDH and 6PGDH activity in, 171, 174; attachment to a surface by, 13; early work in growing of, 16, 17; food distribution paths and rates in, 119; growth and development of, 13–36; growth media for, 22–24; growth rates of, 24 fig 3; hydranth increase rate in, 19 tab 2, 24; laboratory cultures of, 16–19; polyp and medusa stages of, 16; species maintenance in laboratory of, 17, 18 tab 1; stolon growth in, 25–32
Hydroxyproline: activity in *Aiptasia* sp., 148, 149 fig 1, 153, 154; feeding responses of *Cyphastrea ocellina* to, 107 fig 4, 108, 116; in *Hydra littoralis,* 157; in *Physalia physalis,* 157
Hydrozoa, 160 tab 1, 172 tab 1, 228 tab 1

"Instant Ocean," 22, 23; as medium for *Boloceroides* sp., 92, 169; as medium for *Pennaria tiarella,* 85, 86; composition of, 23

Isoleucine, feeding response of *Boloceroides* sp. to, 95 tab 1, 96, 98
Isorhiza, 163, 164; apotrichous, 164, 166 fig 7; atrichous, 163, 166 fig 4; basitrichous, 163, 166 fig 5; holotrichous, 164, 166 fig 8; merotrichous, 163, 166 fig 6

Larvae, planular, of *Pocillopora damicornis,* 66, 67 fig 1, 68–70
Leptastrea bottae, 49, 115; feeding responses of, 112, 113 tab 6
Leptastrea purpurea, 49
Leptomedusae (= Calyptoblastea), 18 tab 1
Leucine, feeding response of *Boloceroides* sp. to, 95 tab 1, 96, 98
Lophelia obtusa, skeletal mineralogy of, 231
Lovenella (= *Eucheilata*) *clausa,* 18 tab 1
Lumbricus sp., as diet supplement for *Cordylophora lacustris,* 20

Macranthea cookei, 45, 47 fig 12, 49, 149; effect of DTE upon, 161 tab 1; G6PDH and 6PGDH activity in, 173 tab 1
Madreporaria (Scleractinia), 49, 161 tab 1, 173 tab 1, 228 tab 1
Magnesium ions: in *Chlorohydra viridissima,* 10; in *Cordylophora lacustris,* 10; in *Hydra littoralis,* 10
Manicina areolata, 185, 232
Margelopsis haeckeli, 17 tab 1
Mastigias papua, 49, 50 fig 14
Mastigophores, 164; macrobasic, 164, 166 fig 12; microbasic, 164, 166 fig 11
Merga galleri, 17 tab 1
Mestigius sp., 174 tab 2; G6PDH and 6PGDH activity in, 172 tab 1
Methionine, 89
Metridium marginatum, 159

Index

Metridium senile: mesoglea of, 146, 149; proline-hydroxyproline ratio in, 154

Milleporina, 228 tab 1

Mitosis: in *Campanularia flexuosa,* 27, 28; in *Clytia johnstoni,* 27, 28; in colonial hydroids, 27, 28; in *Cordylophora lacustris,* 28; effect of temperature on, 31, 32; in *Proboscidactyla* sp., 28

Mitrocomella (= *Cuspidella*) *brownei,* 18 tab 1

Montastrea annularis, 185

Montipora sp. (perhaps *M. verrilli*): amino acid analysis of, 261; analysis of organic matrix of, 260; chromatography of, 261; collection and maintenance of, 260; hydrolysis of matrix of, 261, 262 tab 1, 263; isolation of matrix of, 261, 263, 264

Montipora flabellata, 49

Montipora verrucosa, 49; feeding on labeled bacteria by, 133

Myohaematins. *See* Cytochrome

Mytilus sp., as food supplement for hydras, 21

Nanomia cara, feeding responses of, 77 tab 1

Nematocyst, control of discharge in hydras, 9

Nemopsis dofleini, 17 tab 1; food requirements of, 21, 22

Nitrate ions: in *Cordylophora lacustris,* 10; in *Hydra littoralis,* 10

Obelia sp., 18 tab 1, 174 tab 2; G6PDH and 6PGDH activity in, 172 tab 1

Obelia geniculata, 27; stolon growth rate of, 25

Oulastrea, 181, 184

Pachydrilus lineatus, as diet supplement for *Cordylophora lacustris,* 20

Palythoa sp., 47, 209; G6PDH and 6PGDH activity in, 173 tab 1

Paramecium bursaria, 180–182

Pavona explanulata, 49

Pavona varians, 49; amino acid analysis of, 261; analysis of organic matrix of, 260; chromatography of, 261; collection and maintenance of, 260; feeding on labeled bacteria by, 133; hydrolysis of matrix of, 261, 263; isolation of matrix of, 260, 261, 263, 264

Pelagia noctiluca, 154

Pennaria tiarella: autoradiography of, 120, 123, 124 fig 2; collection and preparation of, 119, 120; description of, 84 fig 1, 86 fig 2, 121 fig 1; distribution of radioactivity in, 121, 122 tab 2, 123, 124 fig 2, 125 tab 4, 127, 128; effect of DTE upon, 159, 160 tab 1; feeding behavior of, 84–91; feeding of radioactive food to, 120, 123, 127; feeding responses of, 77 tab 1; feeding response to *Artemia salina,* 87, 88 fig 3, 91, 127; feeding response to pipecolic acid of, 88, 89, 107; feeding response to proline of, 88, 89 tabs 2–3, 97–100, 107, 114; G6PDH and 6PGDH activity in, 171, 172 tab 1; hydranth response of, 86; "Instant Ocean" as medium for, 85, 86; mouth-spreading response of, 88 tab 1; protein determination of, 120, 121 tab 1; retention of radioactivity in, 124, 125 tab 3, 127

Pennariidae, 48

Pennatulacea, 228 tab 1

Physalia physalis: effect of glutathione on, 92; feeding responses of, 77 tab 1; G6PDH and 6PGDH activity in, 171, 172 tab 1; mesoglea of, 146; nematocyst capsules of, 157; proline-hydroxyproline ratio in, 154

Physalia utriculus, 49, 172 tab 1;

mouth-spreading response in, 88;
6PGDH activity in, 171
Pipecolic acid: as feeding activator for
Cordylophora lacustris and *Pennaria
tiarella,* 88, 89 tab 2, 107; as
mouth-opening agent in *Cyphastrea
ocellina,* 100, 107, 114, 116
Plumularia sp., G6PDH and 6PGDH
activity in, 172 tab 1
Pocillopora bulbosa, expulsion periods
of planulae in, 71
Pocillopora damicornis, 42 fig 6, 43 fig
7, 49, 115; aberrant planular larvae
of, 70 fig 5, 71 fig 6; amino acid
analysis of, 261; analysis of organic
matrix of, 260, 264; *Artemia salina*
as food for, 67, 68, 112, 248;
chromatography of, 131, 161;
collection and maintenance of, 66,
67, 129, 130, 246, 260; feeding
of labeled bacteria to, 130, 131, 132
tab 1, 133, 134 tab 2, 135; feeding
of planular larvae of, 67, 69; feeding
responses of, 112 tab 5, 113 tab 6,
115, 116, 132; fractionation of, 131,
135; growth of larvae of, 68 figs 2–3,
69 fig 4; G6PDH and 6PGDH activity
in, 173 tab 1; hydrolysis of matrix
of, 261, 262 tab 1; isolation of
matrix of, 260, 264; matrix analysis
of, 231, 232; radioactivity in, 249,
251; skeleton of calyx in, 230, 231;
temperature experiments on, 248,
249, 251, 252 fig 4, 254 fig 6,
255–259; uptake of ^{45}C by, 246,
248, 250 tab 1, 251, 252, 253 fig 5,
256; weight correlations in, 246,
247, 248 fig 1, 249 fig 2, 250 fig 3
Pocillopora meandrina, 43 fig 8, 45, 49,
115; effect of DTE on, 161 tab 1;
feeding on labeled bacteria by, 133;
feeding responses of, 112, 113 tab 6
Podarke, 63
Podocoryne carnea: influence of
quantity of food upon, 22, 24 fig 4;
laboratory culture of, 9, 17 tab 1
Podocoryne hartlaubi, 17 tab 1; food
requirements of, 21, 22

Porites sp. (perhaps *P. compressa*):
amino acid analysis of, 261; analysis
of organic matrix of, 260;
chromatography of, 261; collection
and maintenance of, 260; hydrolysis
of matrix of, 261, 262 tab 1, 263;
isolation of matrix of, 261, 263, 264
Porites compressa, 41, 49, 115, 132,
159; effect of DTE on, 161 tab 1;
feeding on labeled bacteria by, 133;
feeding responses of, 112, 113 tab 6
Porites lobata, 41, 43, 49
Poritidae, 41–43, 49
Potassium ions: in *Chlorohydra
viridissima,* 10; in *Hydra littoralis,*
10
Proboscidactyla sp., mitosis in, 28
Proline: activity of, in *Aiptasia* sp., 148,
149 fig 1, 153, 154; feeding response
of *Cordylophora lacustris* to 88, 89,
91, 92, 97, 100; feeding response of
Cyphastrea ocellina to, 90, 97, 100,
104 fig 2, 105, 107, 109, 114–116;
feeding response of *Pennaria tiarella*
to, 88, 89 tabs 2–3, 90, 92, 97, 100
Protohydra leuckarti, 17 tab 1
Psammacora stellata, 49

Rathkia octopunctata, 17 tab 1
Rhabdoides, 164
Rhizorhagium album, 17 tab 1
Rhizostomaceae, 49
Rhopaloides, 164
Rhopalonemes, 163

Sagitta sp., as food supplement for
hydras, 22
Sarsia tubulosa, 17 tab 1
Schmidt-Thanhauser technique, 219
Scleractinia, 37–45
Scyphozoa, 160 tab 1, 172 tab 1
Seriatopora caliendrum, 232
Seriatoporidae, 43–45, 49
Siphonophora, 49, 172 tab 1
6PGDH. *See* 6-phosphogluconate
dehydrogenase
6-phosphogluconate dehydrogenase
(6PGDH), 169–171, 172 tab 1,

173-175; fluorometric measurements of, 170
Skeletonema costatum, 129
S-methyl glutathione, feeding responses of *Cyphastrea ocellina* to, 108 fig 5, 114, 116
Sodium ions: in *Chlorohydra viridissima,* 10; in *Hydra littoralis,* 10
Spirocysts, 165, 167 fig 23
Spongilla sp., 180, 181
Stauridiosarsia japonica, 17 tab 1; food requirements of, 21, 22
Staurocladia portmanni, 17 tab 1; influence of quantity of food upon, 22
Staurocoryne filiformis, 17 tab 1
Stenoteles, 165, 167 fig 21
Stolon, 27, 28; ectoderm cell movement in, 29, 30; regions of, 28, 29
Stolon, growth of: in *Campanularia flexuosa,* 25; cell division in, 28; in *Clava multicornis,* 25; in colonial hydroids, 19 tab 2, 24-27; in *Cordylophora lacustris,* 25, 26; in *Hydractinia echinata,* 27; in *Obelia geniculata,* 25; patterns of, 28; rhythmic cycles of, 30, 31 fig 5; in *Tabularia crocea,* 26
Stolonal elongation: activity of contractile tissues of, 29; in colonial hydroids, 27-32
Stolonifera, 228 tab 1
Stomocnides, 163, 164
Stylasterina, 228 tab 1
Syncoryne exima: basic requirements of culture of, 18; devices for growing, 18, 19 fig 1, 20 fig 2; early experiments in growth of, 17 tab 1, 18, 19 tab 2

Telestacea, 228 tab 1
Thamnasteriidae, 49
^{35}S-labeled amino acids, 222; feeding to *Aiptasia* sp. of, 218-220; fractionation of, 219, 221, 222 tabs 1-2
Trachylina, 18 tab 1

Tridacna sp., 186
Tubastrea aurea, 44 figs 9-10, 45, 49
Tubastrea manni, 115; effect of DTE on, 161 fig 1; feeding responses of, 112, 113 fig 6
Tubularia crocea, 18 tab 1; stolonal growth rate in, 26
Tyrosine, 127

Valine: as feeding activator of *Boloceroides* sp., 89, 95 tab 1, 96-98, 100
Velella pacifica, 49

Xeniidae, 187

Zanclea implexa, 18 tab 1
Zoanthidea, 47-49, 161, 173 tab 1, 187
Zoanthus sp., 47, 48 fig 13, 49, 185; G6PDH and 6PGDH activity in, 173 tab 1
Zoanthus sandwichensis: Artemia salina unacceptable as food for, 209; bacterial analysis of, 211; chromatography of, 210; collection and maintenance of, 209; effect of DTE on, 161 tab 1; feeding habits of, 209; ^{14}C distribution in, 212 tab 2; fractionation of, 210; incubation with ^{14}C-glycine of, 209; radioactivity in, 210; uptake of ^{14}C-glycine of, 209, 211 tab 1, 213, 214, 215 tab 3
Zoochlorellae: of algae, 179, 180; of coelenterates, 179
Zooxanthellae, 235; calcification of, 235, 236; coral reef productivity of, 187, 188; digestion of, 184; in sea anemones, 184; method of assay of, 219; of algae, 179, 180; of coelenterates, 181-183, 188; photosynthesis by, 184-187; role in nutrition of host of, 183, 184; uptake of ^{35}S by, 220, 221 fig 1, 222 tab 1